Supervisory Control and Scheduling of Resource Allocation Systems

Supervisory Control and Scheduling of Resource Allocation Systems

Reachability Graph Perspective

Bo Huang
Nanjing University of Science and Technology, Nanjing, China

MengChu Zhou
New Jersey Institute of Technology, Newark, NJ, USA

IEEE Press Series on Systems Science and Engineering
MengChu Zhou, Series Editor

WILEY

Published by John Wiley & Sons, Inc., Hoboken, New Jersey.
Published simultaneously in Canada.

For general information on our other products and services please contact our Customer Care Department with the U.S. at 877-762-2974, outside the U.S. at 317-572-3993 or fax 317-572-4002.

Wiley also publishes its books in a variety of electronic formats. Some content that appears in print, however, may not be available in electronic format.

Library of Congress Cataloging-in-Publication Data

Names: Huang, Bo, 1980- editor. | Zhou, MengChu, editor.
Title: Supervisory control and scheduling of resource allocation systems : reachability graph perspective / [edited by] Bo Huang, Nanjing University of Science and Technology, Nanjing, China, Mengchu Zhou New Jersey Institute of Technology, Newark, NJ, USA.
Description: First edition. | Hoboken, New Jersey : John Wiley & Sons, Inc., [2020] | Series: IEEE Press series on systems science and engineering | Includes bibliographical references and index.
Identifiers: LCCN 2020016184 (print) | LCCN 2020016185 (ebook) | ISBN 9781119619680 (cloth) | ISBN 9781119619697 (adobe pdf) | ISBN 9781119619703 (epub)
Subjects: LCSH: Resource allocation–Decision making. | Management information systems.
Classification: LCC T57.77 .S87 2020 (print) | LCC T57.77 (ebook) | DDC 658.4/034–dc23
LC record available at https://lccn.loc.gov/2020016184
LC ebook record available at https://lccn.loc.gov/2020016185

Cover Design: Wiley
Cover Image: © Mark Agnor/Shutterstock

Set in 9.5/12.5pt STIXTwoText by SPi Global, Pondicherry, India

10 9 8 7 6 5 4 3 2 1

Contents

Preface

This book covers the supervisory control and scheduling of resource allocation systems (RASs) based on the reachability graphs of Petri nets that allow for multiple resource acquisitions and flexible routings. In such systems, resources (such as machines, robots, drives, and data) are shared among concurrently running processes (such as parts, vehicles, and programs). A successful system should correctly allocate these resources without deadlocks and at the same time achieve some desired objectives such as maximizing makespan and minimizing tardiness.

First, deadlocks in such systems can lead to an unnecessary economic cost and even catastrophic results in practical systems. In the literature, Petri nets (PNs) are widely used to describe and control RASs since PNs can naturally and compactly model the structural and behavioral properties of RASs and then analyze the performances of the models. To obtain deadlock-free supervisors for PN models of RASs, a reachability graph analysis method, which synthesizes a PN supervisor by using an invariant-based control approach, is commonly used. This method can be applied to many kinds of PN models and allows designers to obtain a deadlock-free supervisor with maximally or highly permissive behavior, but its computational burden is always heavy and of exponential complexity. In the first half of this book, the reachability graph analysis method for deadlock prevention of RASs are accelerated by using different strategies.

Second, although RASs allow high productivity, a great number of choices for resources and processing routes make it difficult to correctly allocate different resources to different processes and to efficiently schedule the sequence of operations with the highest efficiency. In order to address these problems, an A* search with an admissible heuristic function that only needs to explore a partial reachability graph to obtain an optimal schedule can be used. Although the generation of the whole reachability graph is avoided, the number of the explored markings by the algorithm still grows exponentially with problem size and/or initial markings, which makes the method only applicable to small systems. However, result quality and computational efficiency are two important factors in the scheduling of RASs.

In the second half of this book, different acceleration strategies for the A* search within reachability graphs, such as hybrid searches, more informed heuristics, and symbolic representations and manipulations, are presented in order to speed up the search process for optimal or suboptimal schedules.

A lot of work has been done by researchers and engineers in both the deadlock handling and system scheduling of RASs. This book focuses on the solutions' quality and computational speed in RAS deadlock handling and system scheduling based on Petri nets and their reachability graphs. If optimal or suboptimal solutions can be obtained with fewer computational burdens, more systems can be handled by the deadlock prevention and system scheduling methods. Therefore, this book aims to present the state-of-the-art developments in the acceleration of the supervisory control and scheduling of RASs based on PNs' reachability graphs. The outline of this book is as described below:

The book is divided into three parts. Part I includes Chapters 1 and 2 that introduce the basic concepts and research background of RASs and Petri nets. Part II consists of Chapters 3–7 focusing on the deadlock-free supervisor synthesis for RASs by using Petri nets. Part III contains Chapters 8–12 and investigates the heuristic scheduling of RASs based on timed Petri nets.

Chapter 1 introduces the features of RASs and briefly reviews different approaches on the supervisory control and system scheduling of such systems. Then, the advantages and disadvantages of these approaches are analyzed.

Chapter 2 first recalls the fundamentals of Petri nets, including their basic concepts, modeling power, behavioral properties, and analysis tools. Next, some typical subclasses of Petri nets for RASs and their relationships are introduced. After that, structural analysis methods and reachability graph analysis methods based on PNs for RASs are discussed. Finally, the procedures and properties of the heuristic A* search within the PNs' reachability graphs are presented. The basic concepts given in this chapter are used throughout the book.

Chapter 3 focuses on the applications of the theory of regions in the synthesis of optimal and deadlock-free supervisors with the fewest monitors for RASs based on PN's reachability graphs. The performance of such a deadlock prevention policy can be evaluated by three criteria: behavioral permissiveness, structural complexity, and computational complexity. This chapter presents a supervisor synthesis method that has the following advantages: its controlled nets are behaviorally maximal and the added supervisors are structurally minimal in terms of added monitors. However, its computational burden is usually heavy.

The existing reachability graph analysis methods for deadlock prevention consider all activity places in the design of place invariants that prevent the system from entering the deadlock zone in the graph. However, some activity places may be useless for such a method and they may result in many redundant constraints in the supervisor synthesis process. To handle it, Chapter 4 presents a method to obtain liveness-enforcing and optimal or suboptimal supervisors for RASs based on critical activity places that are a proper subset of activity places. The proof that critical activity places are enough to be considered in the design of place invariants to obtain such supervisors is given. Since the number of places considered in the supervisor synthesis is reduced, the supervisor synthesis process thus becomes simpler.

To reduce the computational burden of the method in Chapter 3, researchers have mainly focused on the revision of the underlying integer linear program (ILP). Instead, Chapter 5 presents another strategy to accelerate the computation by eliminating redundant reachability constraints given in the method. First, a sufficient and necessary condition for a reachability constraint to be redundant is given in the form of an ILP, based on the concept of the feasible region of supervisors. Then, two kinds of redundancy elimination methods are presented: an ILP method and a non-ILP method. Most of the redundant reachability constraints can be eliminated by the presented methods in a short time. Thus, the computational time of the deadlock prevention method is greatly reduced after the elimination, especially for large-scale systems. In addition, the obtained supervisors are still optimal and structurally minimal.

Chapter 6 investigates an acceleration method for solving a multiobjective ILP to obtain an optimal and deadlock-free supervisor with a compressed structure in terms of both control places and added arcs for RASs. Instead of a single big ILP, several smaller ILPs are formulated in an iterative manner and the problem can thus be solved much faster. To further reduce the ILP solution time, the redundancy elimination method is also adopted in the iterative method. The advantages of the presented method are that the supervisor synthesis process is well accelerated without compromising the result's optimality and the obtained supervisor is structurally compressed in terms of both control places and added arcs.

Chapter 7 presents a deadlock prevention method for RASs based on Petri nets with uncontrollable and unobservable transitions. First, the concepts of admissible markings and first-met inadmissible markings are introduced. Next, place invariants are designed to survive all admissible markings and prohibit all first-met inadmissible markings, which keeps the system from reaching deadlock markings, livelock markings, bad markings, and those that can evolve into deadlocks, livelocks, or bad markings via the firings of uncontrollable transitions. Then, an ILP is developed to ensure that the obtained supervisor does not violate the unobservability and the supervisor is structurally minimal in terms of control places and added arcs. The presented method can deal with the PNs in which some crucial transitions are uncontrollable and/or unobservable. The supervisors obtained by the method are admissible. In addition, the condition under which the obtained supervisors are optimal is also given.

Chapter 8 first presents two methods to build place-timed Petri net models for the RAS scheduling, i.e. converting existing Petri nets of RASs in the literature and synthesizing a new one via a top-down or a bottom-up method. Then, the state evolution of such place-timed Petri nets is introduced in detail. Finally, an algorithm that combines the evolution of a place-timed Petri net, an intelligent A* search, and state check rules is presented to schedule RASs.

Chapter 9 provides a good trade-off between the quality of the obtained schedule and the computational burden of the RAS scheduling based on PN's reachability graphs. This chapter presents an admissible heuristic function for the place-timed Petri nets of RASs and two controllable search methods with the heuristic function. The first method does not need to predict the depth of a solution in advance and the quality of the obtained schedule is controllable. The second method combines an A* search with a depth-first strategy within the PN's reachability graphs to schedule RASs. The search scheme can invoke quicker termination conditions and the quality of the obtained result is controllable.

Chapter 10 presents a hybrid scheduling method that combines two search schemes for RASs based on place-timed PNs. The method performs an A* algorithm locally and a backtracking strategy globally within the PNs' reachability graphs. It is proven to be more efficient than other existing methods that also combine the A* algorithm and the backtracking strategy.

Chapter 11 deals with the method that can simultaneously use admissible and inadmissible heuristic functions in an A* search within the reachability graphs of PNs for RASs. It also proves that the combinational heuristic function is still admissible and more informed than any of its constituents.

Chapter 12 investigates a symbolic A* search to compactly represent and efficiently schedule RASs based on reachability graphs. First, the methods to functionally represent, evolve, and schedule the place-timed PNs by using both binary decision diagrams and an A* algorithm are presented. To accelerate the search process, an admissible heuristic function suitable for the symbolic method and its efficient functional implementation are given. When compared with explicit-state A* methods, the presented symbolic A* approach can compactly represent large sets of states and efficiently compute on such sets to fast find an optimal schedule.

Chapter 13 provides several interesting open problems and future research topics in the development of deadlock prevention and system scheduling methods for RASs based on Petri nets.

Attached to the end of the book is a reference bibliography. A glossary and a list of acronyms can be found at the beginning of the book and a complete index is provided at the end of the book. The implementation codes and experimental files of the methods presented in this book are also given for the reader's examination and reference. For the codes and input data of the methods presented in Part II, please refer to http://github.com/huang-njust/supervisor. The programs and experimental data of the methods in Part III except Chapter 12 can be seen in http://github.com/huang-njust/schedule. The codes and input data of the method presented in Chapter 12 are given in http://github.com/huang-njust/bdd.

This book can be used as a reference for university classes aimed at modeling, analysis, control, and scheduling systems based on discrete event models. The book can provide useful information for control engineers and industrial engineers who are interested in developing deadlock prevention and schedule optimization methods for RASs. It may also be of interest to the discrete event systems community, providing a bridge for supervisory control and system scheduling applications. The book serves researchers, engineers, scientists, and professionals who work in system modeling, planning, control, engineering, and management fields as well as faculty, staff, and graduate students who are interested in Petri net, scheduling, and control of discrete event dynamic systems.

Bo Huang and MengChu Zhou
Nanjing, China

Acknowledgments

We cannot acknowledge all the people who have made suggestions and given help, but we would like to sincerely note the help from Professors Yufeng Chen, Macau University of Science and Technology (Macau, China), Yan Qiao, Macau University of Science and Technology (Macau, China), Yisheng Huang, Taiwan ILan University (Taiwan, China), Zhiwu Li, Macau University of Science and Technology (Macau, China), Naiqi Wu, Macau University of Science and Technology (Macau, China), Maria Pia Fanti, Polytechnic University of Bari (Italy), Shouguang Wang, Zhejiang Gongshang University (China), Dmitry Zaitsev, International Humanitarian University (Ukraine), Jiliang Luo, Huaqiao University (China), Jun Li, Southeast University (China), Qinghua Zhu, Guandong University of Technology (China), Chunrong Pan, Jiangxi University of Science and Technology (China), Peiyun Zhang, Anhui Normal University (China), Yisheng An, Changan University (China), and Juan-Pablo López-Grao, University of Zaragoza (Spain), Dr. Huanxin Henry Xiong, Nokia (USA), and the esteemed anonymous reviewers of this book. We would like to appreciate the assistance provided by some of the first author's graduate students from Nanjing University of Science and Technology (NUST), who helped to prepare some diagrams of this book.

The first author would like to thank his doctoral thesis advisor Professor Yamin Sun, NUST, who introduced him to this exciting research field. He is also very thankful to his NUST colleagues for their great professional help and friendship. Finally, he appreciates his wife, Liwei Luo, and his lovely daughter, Kexin Huang, for their support during the long period of writing this book.

The second author would like to thank many of his mentors including Professors Frank DiCesare, James Tien, Peter Luh, Frank Lewis, Keith Hipel, Tsu-Tian Lee, and Ljiljana Trajkovic, Dr. Dimitar Filev, and Dr. Edward Tunstel. He would also thank many colleagues at New Jersey Institute of Technology. Finally, he would thank his family members and many friends.

This monograph was in part supported by the National Natural Science Foundation of China under Grants 61773206 and 61203173, the Natural Science Foundation of Jiangsu Province of China under Grant BK20170131, the Foundation of Fujian Engineering Research Center of Motor Control and System Optimal Schedule under Grant FERC002, and the Jiangsu Overseas Visiting Scholar Program for University Prominent Young & Middle-aged Teachers and Presidents.

Bo Huang and MengChu Zhou

Glossary

$\emptyset$	The empty set
$\cap, \cup, \backslash$	Set intersection, union, and difference
$\oplus$	The logical operation of exclusive OR
α	A symbolic function to add one
β	A coefficient of a place invariant
Γ	A positive integer that is big enough
$\delta^{t^j}(p)$	A transformation function of the $(j + 1)$th Boolean variable of $M(p)$ when t fires
η	A symbolic function to subtract one
κ	The tight upper bound of capacities of activity places
λ	The empty string
σ	A sequence of transitions
$\vec{\sigma}$	The Parikh vector of σ
$\tau_{p_{max}}$	The sum of operation times spent on $H_l(p_{max})$ in a state transfer
Φ	The set of marking/transition separation instances
Ω	A feasible region
A_K	The set of critical activity places
A_F	The set of free activity places
$A_{\vec{F}}$	The set of forward free activity places
$A_{\overleftarrow{F}}$	The set of backward free activity places
c	The cost or makespan of a path
c^*	The cost of an optimal path
$c(S, S')$	The cost of a transfer from S to S' in $G(\mathcal{N}, S_0)$
$c^*(S, S')$	The optimal cost of a transfer from S to S' in $G(\mathcal{N}, S_0)$
C	A control vector indicating which transition is to fire
$C(t_j)$	The element of C with respect to transition t_j
d	The depth of a state
D	The vector of time delays in places
$D(p_i)$	The time delay of a place p_i

e	The relative error of a heuristic function
e^+	The positive relative error of a heuristic function
e^-	The negative relative error of a heuristic function
$\mathcal{E}_t$	The enable function of a transition t
f	An estimate of the lowest cost from M_0 (or S_0) to M_G (or S_G) along a path through M (or S)
f^*	The lowest cost from M_0 (or S_0) to M_G (or S_G) among all paths through M (or S)
$f_{j,k}$	An integer variable in $\{0,1\}$ denoting whether or not a place invariant designed to forbid M_j forbids M_k
F	The set of arcs in a Petri net
g	The lowest cost currently found from M_0 (or S_0) to M (or S)
g^*	The lowest cost among all paths from M_0 (or S_0) to M (or S)
$G(N, M_0)$	The reachability graph of a net system (N, M_0)
$G(\mathcal{N}, S_0)$	The reachability graph of a place-timed net system $(\mathcal{N}, S_0)$
h	A heuristic estimate of the cost from M (or S) to M_G (or S_G) along an optimal path
h^*	The cheapest cost from M (or S) to M_G (or S_G)
$H(r)$	The set of holders of r
$H_l(r)$	The set of loyal holders of r
I	A loading buffer
I	A place invariant
$\|I\|$	The support of I
$\mathbf{1}_n$	The $n \times n$ identity matrix
$l_{j,i}$	The ith coefficient of a place invariant I_j
M	A marking
M_0	An initial marking
M_G	A goal marking
$M(p_i)$	The number of tokens in a place p_i at a marking M
$\mathcal{M}$	A set of markings
$\mathcal{M}_A$	The set of admissible markings
$\mathcal{M}_{\tilde{A}}$	The set of inadmissible markings
$\mathcal{M}_{\tilde{C}}^D$	The set of uncontrollable dangerous markings
$\mathcal{M}_D$	The set of markings in the deadlock zone
$\mathcal{M}_L$	The set of legal markings
$\mathcal{M}_L^*$	A minimal covering set of $\mathcal{M}_L$
$\mathcal{M}_L^K$	A minimal covering set of $\mathcal{M}_L$ with respect to K-cover
$\mathcal{M}_F$	The set of first-met bad markings
$\mathcal{M}_F^*$	A minimal covered set of $\mathcal{M}_F$
$\mathcal{M}_F^K$	A minimal covered set of $\mathcal{M}_F$ with respect to K-cover
$\mathcal{M}_{\text{FIM}}$	The set of first-met inadmissible markings

N	A Petri net $N = (P, T, F, W)$
N^x	The xth processing subnet of N
N_{BDD}	The number of used BDD nodes
N_E	The number of expanded states
(N, M_0)	A Petri net system
$[N]$	The incidence matrix of N
$[N]^-$	The pre-incidence matrix of N
$[N]^+$	The post-incidence matrix of N
$\mathcal{N}$	A place-timed Petri net $\mathcal{N} = (P, T, F, W, D)$
$(\mathcal{N}, S_0)$	A place-timed Petri net system
$\mathbb{N}$	The set of natural numbers, i.e. $\{0, 1, 2, \ldots\}$
$\mathbb{N}_\kappa$	Denotes $\{1, \ldots, \kappa\}$
$\mathbb{N}_A$	Denotes $\{i \mid p_i \in P_A\}$
$\mathbb{N}_F^*$	Denotes $\{i \mid M_i \in \mathcal{M}_F^*\}$
$\mathbb{N}_F^K$	Denotes $\{i \mid M_i \in \mathcal{M}_F^K\}$
$\mathbb{N}_J$	Denotes $\{x \in \mathbb{Z}^+ \mid N^x$ is the xth processing subnet of $N\}$
$\mathbb{N}_T$	Denotes $\{i \mid t_i \in T\}$
O	An unloading buffer
O	An objective
p	A place in a Petri net
$p^\bullet$	The set of post-transitions of a place p
$^\bullet p$	The set of pre-transitions of a place p
p_c	A control place
$p_{\max}$	A resource place whose sum of the processing times of its loyal holders is the maximal among all places in P_R^1
P	A part type
P	A set of places
P_0	The set of idle places
P_A	The set of activity (operation) places
P_c	The set of control places
$P_{pre}(t_j)$	A set of t_j's pre-places whose processing times are not zero
$P_{post}(t_j)$	A set of t_j's post-places whose processing times are not zero
P_E	The set of end places
P_R	The set of resource places
P_R^1	The set of 1-bounded resource places
P_S	The set of start places
q_j	An integer variable in $\{0, 1\}$ denoting whether or not a place invariant I_j is selected to design a control place
r	A resource place
R	A resource type
R	The tokens' remaining time matrix of a place-timed Petri net

$R_{i,u}$	The remaining time of the uth token in p_i
R_u	The column vector of R with respect to the uth token
R_S	The set of shared resource places
$R_{\tilde{S}}$	The set of unshared resource places
$R(N, M_0)$	The set of reachable markings of (N, M_0)
$R(\mathcal{N}, S_0)$	The set of reachable states of $(\mathcal{N}, S_0)$
$\mathcal{R}_c$	The percentage of lost optimality
$\mathcal{R}_E$	The percentage of the change in search effort
S	A state of $\mathcal{N}$
S_0	An initial state
S_G	A goal state
$\mathcal{S}$	A BDD representing a set of states
t	A transition in a Petri net
$t^\bullet$	The set of post-places of t
$^\bullet t$	The set of pre-places of t
T	A set of transitions
T_C	The set of controllable transitions
$T_{\tilde{C}}$	The set of uncontrollable transitions
$T_{\tilde{C}}^*$	The set of all finite strings of elements in $T_{\tilde{C}}$
T_{cru}	The set of crucial transitions
T_F	The set of free transitions
$T_{\vec{F}}$	The set of forward free transitions
$T_{\overleftarrow{F}}$	The set of backward free transitions
T_K	The set of critical transitions
T_O	The set of observable transitions
$T_{\tilde{O}}$	The set of unobservable transitions
$T_{O\tilde{C}}$	The set of observable but uncontrollable transitions
$\mathcal{T}$	The computational time
$\mathcal{T}_{H_l(p_i)}$	The loyal time of a resource place p_i
U_p	The resource requirements of an activity place $p \in P_A$
w^j	The $(j+1)$th Boolean variable for the weight of an arc
W	The set of weights of arcs
Z	The set of zones separation instances
$\mathbb{Z}$	The set of integers
$\mathbb{Z}^+$	The set of positive integers

Acronyms

AI	Artificial intelligence
AMS	Automated manufacturing system
BDD	Binary decision diagram
BF	Best first
BT	Backtracking
DEDS	Discrete event dynamic system
DF	Depth first
DP-net	Deadlock prevention net
DWA	Dynamic weighted A^*
DZ	Deadlock zone
ELS^{3}PR	Extended system of linear sequential processes with resources
ES^{3}PR	Extended system of simple sequential processes with resources
ESSP	Event/state separation problem
FBM	First-met bad marking
FIM	First-met admissible marking
FMS	Flexible manufacturing system
GLS^{3}PR	System of generalized linear simple sequential processes with resources
GMEC	Generalized mutual exclusion problem
GPU	Graphics processing unit
IDWA	Improved dynamic weighted A^*
ILP	Integer linear program
LMILP	Lexicographic multiobjective integer linear program
LS^{3}PR	System of linear sequential processes with resources
LZ	Live zone
MFFP	Maximal number of forbidden FBMs problem
MMP	Minimal number of monitors problem
MPP	Minimal number of P-semiflows problem
MTSI	Marking/transition separation instance

NS-RAP	Non-sequential resource allocation process
PC^2R	Process competing for conservative resources
PI	Place invariant
PN	Petri net
PPN	Production Petri net
RAS	Resource allocation system
ROBDD	Reduced ordered binary decision diagram
S^2LSPR	System of simple linear sequential processes with resources
S^2U^2T	Supervisor synthesis with uncontrollable and unobservable transitions
S^3PGR^2	System of simple sequential processes with general resource requirements
S^3PMR	System of simple sequential processes with multiple resources
S^3PR	System of simple sequential processes with resources
S^4R	System of sequential systems with shared resources
SC-net	System scheduling net
SCC	Strongly connected component
SCT	Supervisory control theory
SSP	States separation problem
TPN	Timed Petri net
WOT	Weighted operation time
WRT	Weighted resource time
WS^3PR	Weighted system of simple sequential processes with resources
WS^3PSR	Weighted system of simple sequential processes with several resources

About the Authors

Bo Huang received a PhD degree from Nanjing University of Science and Technology, Nanjing, China, in 2006. He was a Visiting Professor at the University of Missouri-Kansas City, MO, USA, in 2013, and a Visiting Scholar at New Jersey Institute of Technology, Newark, NJ, USA, in 2014–2015 and 2019–2020. He is now a professor at Nanjing University of Science and Technology. His research fields include discrete event systems, Petri nets, intelligent automation, intelligent transportation, and multi-robot systems. He has published 60+ papers and presided 10+ projects in these areas.

MengChu Zhou is a Distinguished Professor of Electrical and Computer Engineering at New Jersey Institute of Technology. He made many contributions to the areas of Petri nets, intelligent automation, Internet of things, big data, web services, and intelligent transportation. He has over 800 publications including 12 books, over 500 journal papers, 23 patents, and 29 book chapters. He is the Editor-in-Chief of IEEE/CAA *Journal of Automatica Sinica* and a recipient of Humboldt Research Award for US Senior Scientists from Alexander von Humboldt Foundation, and Franklin V. Taylor Memorial Award and Norbert Wiener Award from IEEE Systems, Man and Cybernetics Society. He is a Fellow of IEEE, IFAC, AAAS, and CAA.

Part I

Resource Allocation Systems and Petri Nets

1

Introduction

This chapter introduces the definitions of resource allocation systems (RASs) and their deadlock resolution and system scheduling. Different approaches to control and schedule RASs based on Petri nets are briefly reviewed. In addition, both the pros and cons of these approaches are introduced to show the development of deadlock resolution and system scheduling methods for RASs based on Petri nets.

1.1 Resource Allocation Systems

Resource scarcity is a common scenario in many discrete event dynamic systems (DEDSs). In such systems, available resources (such as machines, robots, drives, and programs) have to be shared among concurrently running processes (such as parts, vehicles, and data) and they must compete in order to be granted their allocation and achieve some system objectives such as maximizing makespan and minimizing tardiness. These systems are named resource allocation systems (RASs). This chapter briefly introduces different deadlock resolution and scheduling methodologies based on Petri nets, which are in the view of resource allocation and have successfully been applied to many RASs, such as flexible manufacturing systems (FMSs) (Li *et al.*, 2008*b*), project management systems (Kumar & Ganesh, 1998), and multithread software engineering systems (López-Grao & Colom, 2012).

According to Reveliotis *et al.* (1997), an RAS usually contains a set of resource types $R_i, i = 1, \ldots, |R|$, and a set of jobs $P_j, j = 1, \ldots, |P|$. Each resource type R_i is further characterized by its capacity that is a finite positive integer indicating at most how many units of R_i are in the system. A job P_j often contains a set of tasks or job stages $P_{jk}, k = 1, \ldots, |P_j|$, representing the operations to be processed to finish the job. The operational assumptions on job routing structures and resource requirements give rise to complex behavioral patterns, leading to different control and scheduling approaches. In terms of the resource requirements for jobs, RASs

Supervisory Control and Scheduling of Resource Allocation Systems: Reachability Graph Perspective, First Edition. Bo Huang and MengChu Zhou.

can be categorized into the following four classes: (i) single-unit RASs in which each operation only requires one resource from a single resource type; (ii) single-type RASs in which each operation requires an arbitrary number of resources from the same resource type; (iii) conjunctive (AND) RASs in which each operation requires an arbitrary number of resources from different resource types; and (iv) disjunctive/conjunctive (AND/OR) RASs in which an operation may be associated with a set of conjunctive requests for different resource types. Among them, the AND/OR RASs are the most generalized ones.

In an RAS, there are often a limited number of shared resources to be allocated. The competition for such shared resources by different operations may cause deadlocks where two or more operations are each indefinitely waiting for the other to release their acquired resources. Deadlocks can lead to unnecessary economic cost and even catastrophic results in practical systems. Hence, they are a highly undesirable situation.

For a deadlock in RASs to occur, four conditions must be satisfied (Coffman *et al.*, 1971):

1) Mutual exclusion: when operations claim exclusive control of the resources they need;
2) Hold and wait: when operations hold resources already allocated to them while waiting for additional resources.
3) No preemption: when a resource can only be voluntarily released by the operation holding it.
4) Circular wait: when a circular chain of operations exists, where each operation holds one or more resources that are being requested by the next operation in the chain.

If a deadlock occurs, the above four conditions are met. To handle deadlocks in RASs, most approaches work by preventing one of the four conditions from being satisfied. In general, there are four deadlock handling approaches: deadlock ignoring, deadlock detection and recovery, deadlock avoidance, and deadlock prevention.

Deadlock ignoring, which is also known as an ostrich algorithm, assumes that deadlocks are exceedingly rare in the system and the influence incurred by a deadlock is tolerable. In this case, deadlock ignoring is acceptable from a technical and economical point of view. For example, in UNIX, if the process table is full and processes still try to fork some subprocesses, a deadlock can occur. However, it happens rarely and the procedures to prevent it are cumbersome. Thus, it is often ignored.

Deadlock detection and recovery (Reveliotis, 2000; Wysk *et al.*, 1994) uses a monitor to detect and recover from deadlocks. When a deadlock occurs, the monitor detects it and then actions are taken to either terminate operations in the deadlock or release some resources held by operations to recover the system from the deadlock. This approach requires the periodic detection of deadlocks, which uses a large amount of data and may become complex if several kinds of shared resources are considered. Its efficiency depends on the response time of the implemented algorithms for deadlock detection and recovery. Thus, it is only applied to systems where detection and recovery strategies are not expensive.

Deadlock avoidance (Ezpeleta & Recalde, 2004; Fanti *et al.*, 2006; Hsieh, 2004; Lawley, 2000; Luo *et al.*, 2019; Reveliotis, 2007; Wu & Zhou, 2005; Xing *et al.*, 2009) is a dynamic and online approach that works in a real-time manner and bases the decision on information about the resource allocation status. In deadlock avoidance, both deadlocks and bad states (which are the states that inevitably lead to deadlocks) are avoided by using look-ahead procedures. The approach usually has high resource utilization and throughput since it tries to permit more states in a system, but sometimes does not eliminate all deadlocks. If it cannot avoid all deadlocks, some recovery strategies are required (Wysk *et al.*, 1994).

Deadlock prevention (Abdallah *et al.*, 2002; Chen *et al.*, 2016; Ezpeleta *et al.*, 1995; Fanti & Zhou, 2004; Feng *et al.*, 2020; Li & Zhao, 2008; Luo *et al.*, 2009; Tricas *et al.*, 2000; Zhou & DiCesare, 1991) works by preventing at least one of the four conditions necessary for the occurrence of a deadlock at any point of the RAS dynamic. It is an off-line computational approach which controls the requirements for resources to ensure that deadlocks never occur. When compared with deadlock avoidance, this approach tends to be more conservative and reduces resource utilization and system productivity. However, it has the advantages of safety and stability. In addition, its controller implementation does not need the knowledge of system states, which leads to a simple control law.

For deadlock resolution of RASs, there are mainly three modeling tools: digraphs, automata, and Petri nets. Each tool has its advantages and has greatly improved the researchers' insight into deadlock phenomena of RASs. Digraphs or graph theory is a simple and an intuitive tool to describe the interactions between operations and resources in RASs. In a digraph, deadlocks are usually related to the circuits in the graph. Thus, the occurrence of deadlocks can be detected by computing all the circuits of the graph and can then be eliminated by some deadlock resolution policies. In this field, some representative research groups are led by Wysk *et al.* (1991, 1994) and Fanti *et al.* (1997, 2002). Supervisory control theory (SCT), originated by Ramadge and Wonham (1987), uses formal language and finite state automata to provide a formal and generic framework of modeling and control of DEDSs. SCT has a profound influence on the supervisory control of DEDSs by using other formalisms such as Petri nets. Many effective and computationally efficient deadlock control policies are developed based on automata. Reveliotis, Ferreira, and Lawley are the representative researchers in this field (Lawley & Mittenthal, 1999; Lawley *et al.*, 1998; Reveliotis & Ferreira, 1996). Particularly, Reveliotis *et al.* (1997) propose a significant deadlock avoidance strategy with polynomial complexity for a class of RASs based on finite state automata, which is then extended based on Petri nets by Park and Reveliotis (2001).

On the other hand, to meet the needs of market development, many practical RASs have to quickly respond to changes in the market by integrated planning and scheduling, which allocate limited shared resources to different operations and determine the sequences of operations so that the constraints of the system are met and performance criteria are optimized. RASs usually contain a great number of choices of resources and processing routes to allow high productivity. Thus, it imposes a challenging problem, that is, how to correctly allocate shared resources to different operations and effectively schedule the operations to achieve the highest efficiency. Such a scheduling problem is NP-hard and its computational time to obtain an optimal schedule grows exponentially with the problem size (Cao *et al.*, 2019; Kolisch & Drexl, 1997). To handle the RAS scheduling problem, there are generally two kinds of scheduling methods: exact methods and heuristic methods.

Exact methods are useful in obtaining an optimal schedule for small-scale RASs. For example, linear programming is a useful method that integrates various decision variables into one model to get an optimal solution for an RAS scheduling problem (Burkard & Hatzl, 2005; Qiao *et al.*, 2017). It focuses on a single objective such as maximizing the profit and minimizing the cost. However, practical problems do not always have one objective to achieve. For this reason, goal programming that exploits multiobjective optimization becomes attractive for real-world RAS scheduling (Azaiez & Al Sharif, 2005; Moro & Ramos, 1999). In general, such mathematical programming methods can ensure the optimality of the obtained solutions, but they have some limitations over complex systems since it is difficult for them to describe the constraints in complex RASs, such as the RASs with shared resources and alternative routes. In addition, exact methods are usually applied to small systems due to their heavy computational burden.

When exact methods are too slow to obtain optimal schedules for RASs, heuristic algorithms can be used to obtain suboptimal schedules in a reasonable time frame. Many researchers have combined heuristic dispatching rules (Dominic *et al.*, 2004; Li *et al.*, 2013) such as first come first serve, shortest processing time, and least working remaining time, with the evolution of systems' models to handle the RAS scheduling. Dispatching rules are often easy to use in a practical system, but they have limited information about the entire system and fail to schedule complex models well. To address this problem, metaheuristics are usually used. The representative metaheuristic strategies for RAS scheduling in the literature are simulated annealing (Jozefowska *et al.*, 2001; Yuan *et al.*, 2019a), genetic algorithm (Hou *et al.*, 2017; Pezzella *et al.*, 2008), tabu search (Nonobe

& Ibaraki, 2002; Zuo *et al.*, 2019), ant colony optimization (Feng, *et al.*, 2019; Rajendran & Ziegler, 2004), particle swarm optimization (Kang *et al.*, 2016; Sha & Hsu, 2006), and metaheuristic hybridization (Chen & Shahandashti, 2009; Yuan *et al.*, 2019b). The metaheuristic methods can often provide good solutions in a reasonable period of time, but it is rarely possible to tell how close the obtained solutions are to the optimal solution. Therefore, RAS scheduling is a typical combinatorial optimization problem for which a desirable scheduling method should include both easy formulation of the systems and quick computation of optimal or suboptimal solutions.

1.2 Supervisory Control and Scheduling with Petri Nets

Petri nets (Murata, 1989), also known as place/transition nets, are a graphical and mathematical formalism which is widely used to model and analyze RASs such as manufacturing systems, computer and communication networks, and transportation systems, because Petri nets can naturally model the systems' structural properties including sequential execution, concurrency, synchronization, and choice, and effectively analyze the systems' behavioral properties such as boundedness and deadlock.

To deal with the deadlock problem in RASs, there are mainly two kinds of Petri-net-based methods: structural analysis (Ezpeleta *et al.*, 1995; Huang *et al.*, 2001; Li & Zhou, 2006, 2008; Wu & Zhou, 2010; Xing *et al.*, 2011; Ye *et al.*, 2018) and reachability graph analysis (Basile *et al.*, 2013; Chen *et al.*, 2011; Ghaffari *et al.*, 2003; Li *et al.*, 2020; Uzam & Zhou, 2006). The former allows one to derive a control policy via special Petri net structures, e.g., resource-transition circuits and siphons, and the resulting control law is usually computationally efficient. However, its applications are often confined to certain classes of Petri nets and its obtained liveness-enforcing supervisors are usually restricted since some permissive behavior may be excluded in the controlled systems. The latter utilizes the Petri net reachability graph that fully reflects the behavior of the underlying system. Although its computational burden is always heavy, it is usually more general, which means that it can be applied to more kinds of Petri nets and often allows designers to obtain deadlock-free supervisors with maximally or highly permissive behavior.

Therefore, reachability graph analysis is an important deadlock resolution technique. In Ghaffari *et al.* (2003), the theory of regions is used to derive an optimal supervisor if such a supervisor exists. To improve its computational efficiency, Uzam and Zhou (2006) develop an iterative approach to deadlock control. They divide a reachability graph into a live zone (LZ) and a deadlock zone (DZ). A marking in LZ is called a legal marking that can reach the initial marking, while DZ contains deadlock markings, livelock markings, and bad markings that inevitably lead to deadlocks and livelocks. First-met bad markings (FBMs) are those in DZ that are immediately reachable from some markings in LZ. At each iteration, to prevent an FBM from being reached, a control place and its arcs are designed by constructing a place invariant according to a well-established invariant-based control method (Yamalidou *et al.*, 1996). This process is iteratively carried out until all FBMs are forbidden. This method is easy to use. However, it does not consider the optimality of the controlled net and the structural minimality of the obtained supervisor.

In Piroddi *et al.* (2008, 2009), reachability graph analysis is combined with siphon analysis to obtain structure-small supervisors with highly permissive behavior. To obtain optimal and structurally minimal supervisors, Chen and Li (2011) propose a deadlock prevention method based on an integer linear program. A vector covering method is used to reduce the number of markings to be considered in the supervisor synthesis. Then, a liveness-enforcing supervisor is obtained by formulating and solving an integer linear program that forbids all FBMs, permits all legal markings, and minimizes the number of added places if such a supervisor exists. The obtained supervisor is optimal and structurally minimal in terms of places in the supervisor. However, it suffers from the state explosion problem produced by an exponential increase of the number of reachable states as system size increases and its computational burden is heavy since it needs to generate the whole reachability graph and often solve a large integer linear program.

On the other hand, the reachability graph of a Petri net represents the evolution of the state space of the underlying system. Thus, Petri nets are also an ideal and popular tool for the scheduling of RASs (Huang *et al.*, 2014; Luo *et al.*, 2015; Lee & DiCesare, 1994; Moro *et al.*, 2002*b*).

When the Petri net model of an RAS is constructed, a search algorithm can run on its reachability graph to find a schedule from a given initial marking to a given goal marking. However, due to the state explosion problem, generating the entire reachability graph is often difficult even for a moderate-sized Petri net model. To deal with this problem, Lee and DiCesare (1994) combine the modeling of Petri nets with an intelligent A* search to find an optimal or suboptimal firing sequence of transitions within the Petri net's reachability graph, only needing to explore a partial reachability graph with a heuristic function. In addition, if the used heuristic function is admissible, the obtained schedule is guaranteed to be optimal. Some admissible heuristic functions for the A* algorithm have been proposed in (Lee & Lee, 2010; Mejía, 2002; Moro *et al.*, 2002*b*; Xiong & Zhou, 1998). However, a problem observed is that, although the generation of the whole reachability graph is avoided, the number of the explored markings still grows exponentially with the problem size and/or initial markings, making this method only applicable to small systems. To speed up the A* search, researchers have developed several improvements on it such as improved heuristic functions (Elmekkawy & Elmaraghy, 2003; Jeng & Peng, 1999), A* algorithms with a controlled or limited backtracking strategy (Mejía, 2002; Moro *et al.*, 2002*a*; Xiong & Zhou, 1998), hybrid A* algorithms to relax the evaluation scope (Lee & Lee, 2010; Moro *et al.*, 2002*b*), the deadlock-free dynamic window search (Luo *et al.*, 2015), and the iterative deepening A* with backtracking (Baruwa *et al.*, 2015). However, they are either only to optimally schedule small-scale systems or to accelerate the search process without controlling the quality of obtained solutions.

1.3 Summary

Results' quality and computational efficiency are two important factors in the supervisory control and system scheduling of RASs. A lot of work has been done by many researchers and engineers in the deadlock handling and system scheduling of RASs to obtain optimal or suboptimal solutions with smaller computational burden. This monograph focuses on both the solutions' quality and computational speed in the deadlock prevention and system scheduling of RASs by using Petri nets and their reachability graphs. Most of the new methods for the supervisory control and system scheduling with reachability graphs are included in this monograph, such as the theory of regions (Ghaffari *et al.*, 2003), the symbolic representations of Petri nets (Chen *et al.*, 2011; Pastor *et al.*, 2001), and some of our recent research results (Huang *et al.*, 2012*a*, 2014, 2015*b*,c, 2017, 2018*a*, 2019).

1.4 Bibliographical Notes

The survey papers and books regarding the deadlock handling and system scheduling of RASs can be found in the literature. The paper (Fanti & Zhou, 2004) surveys a lot of deadlock control approaches in the literature. Li *et al.* (2008*b*, 2012) review and compare many Petri-net-based deadlock control policies. The scheduling methods based on Petri nets and their applications can be found in Tuncel and Bayhan (2007). Rios and Chaimowicz (2010) gives a survey and classifications of the A* algorithms for the pathfinding. A recent survey about deadlock control policies for automated manufacturing systems with unreliable resources is given in (Du *et al.*, 2020). For the books dealing with supervisory control and heuristic search, please refer to Chen and Li (2013), Edelkamp and Schroedl (2012), and Hrúz and Zhou (2007).

2

Preliminaries

This chapter presents a basic overview of Petri nets, going over their modeling power and behavioral properties. Next, several subclasses of Petri nets proposed for resource allocation systems are introduced. Their relationships and example nets are given. Then, their structural analysis for system control and reachability graph analysis for both deadlock prevention and system scheduling are briefly presented. In addition, this chapter introduces the features and usage of INA (Integrated Net Analyzer), which is a popular Petri net modeling, simulation, and analysis tool. Finally, it reviews the definitions and properties of the A* algorithm, which is an intelligent search method used in this book to find optimal or suboptimal schedules within reachability graphs.

2.1 Introduction

Petri nets originated in the Ph.D. dissertation of Carl Adam Petri on "Communication with automata" in 1962. Since then, Petri nets have gained recognition in the research community of different discrete event dynamic systems (DEDSs) such as manufacturing, communications, robotics, engineering, business, and transportation. When compared with other modeling tools for DEDSs such as graph theory and automata, Petri nets have the following main features:

1) Petri nets are a powerful graphical and mathematical modeling tool and can naturally model, simulate, and analyze DEDSs with a high level of concurrency, parallelism, causal dependency, shared resources, and synchronization.
2) Petri nets provide a compact representation of systems that may have a very large state space. They need not to explicitly represent all states of a dynamic system. Instead, only an initial state needs to be represented and the rest reachable states in the state space can be determined by the rules that govern the system evolution.

Supervisory Control and Scheduling of Resource Allocation Systems: Reachability Graph Perspective, First Edition. Bo Huang and MengChu Zhou.

For the modeling of resource allocation systems (RASs), which are a common and typical kind of DEDSs, a variety of subclasses of Petri nets have been proposed in the literature. This chapter introduces their definitions, properties, and relationships. Based on them, this chapter presents two kinds of analysis approaches, i.e., structural analysis and reachability graph analysis, to be used in Part II for deadlock prevention and Part III for system scheduling of RASs. For the reachability graph analysis method, there exists a state explosion problem, i.e., the number of reachable states of a modeled system increases exponentially with the net size and initial markings. To alleviate it, an intelligent A* algorithm can be used to find optimal or suboptimal schedules without exploring the entire reachability graph. This chapter also presents basic definitions and properties of the A* algorithm, to be used in Part III of this book. In addition, a popular Petri net analysis tool, INA, is also introduced.

2.2 Petri Nets

2.2.1 Basic Concepts

Definition 2.1   *A Petri net is defined as a four-tuple $N = (P, T, F, W)$ in which $P = \{p_1, p_2, \ldots, p_m\}$ and $T = \{t_1, t_2, \ldots, t_n\}$ are respectively finite non-empty sets of places and transitions such that $m, n \in \mathbb{Z}^+$ and $P \cap T = \emptyset$, $F \subseteq (P \times T) \cup (T \times P)$ is the set of directed arcs connecting places and transitions, and $W : (P \times T) \cup (T \times P) \to \mathbb{N}$ is a weight assignment for all arcs.*

Example 2.1   An example Petri net is shown in Figure 2.1 in which $P = \{p_1 - p_8\}$, $T = \{t_1 - t_6\}$, $F = \{(p_1, t_1), (t_1, p_2), (p_2, t_2), (t_2, p_3), (p_3, t_3), (t_3, p_1), (p_6, t_4), (t_4, p_4), (p_4, t_5), (t_5, p_5), (p_5, t_6), (t_6, p_6), (p_7, t_1), (t_2, p_7), (p_7, t_5), (t_6, p_7), (p_8, t_2), (t_3, p_8), (p_8, t_4), \quad (t_5, p_8)\}$, $W(p_1, t_1) = W(t_1, p_2) = W(p_2, t_2) = W(t_2, p_3) = W(p_3, t_3) =$

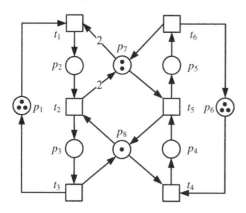

Figure 2.1 A Petri net example.

$W(t_3, p_1) = W(p_6, t_4) = W(t_4, p_4) = W(p_4, t_5) = W(t_5, p_5) = W(p_5, t_6) = W(t_6, p_6) =$
$W(p_7, t_5) = W(t_6, p_7) = W(p_8, t_2) = W(t_3, p_8) = W(p_8, t_4) = W(t_5, p_8) = 1,$ and
$W(p_7, t_1) = W(t_2, p_7) = 2.$ Pictorially, places and transitions are depicted by circles and boxes, respectively. Directed arcs, each of which has a positive number of weight, connect places to transitions, or vice versa but either places to places nor transitions to transitions. Note that if the weight of an arc is one, the arc weight is omitted for conciseness.

Definition 2.2 *For a place $p \in P$, $\bullet p = \{t \in T | (t, p) \in F\}$ is called the set of p's pre-transitions (input ones) and $p^\bullet = \{t \in T | (p, t) \in F\}$ is called the set of p's post-transitions (Output ones). Similarly, for a transition $t \in T$, $\bullet t = \{p \in P | (p, t) \in F\}$ and $t^\bullet = \{p \in P | (t, p) \in F\}$ are called the sets of t's pre-places and post-places, respectively. $\forall x \in P \cup T$, $\bullet\bullet x = \cup_{y \in \bullet x} \bullet y$ and $x^{\bullet\bullet} = \cup_{y \in x^\bullet} y^\bullet$.*

Example 2.2 For the Petri net in Figure 2.1, we have $\bullet p_1 = \{t_3\}$ and $p_1^\bullet = \{t_1\}$. Similarly, $\bullet t_1 = \{p_1, p_7\}$ and $t_1^\bullet = \{p_2\}$. $\bullet\bullet p_1 = \{p_3\}$ and $p_1^{\bullet\bullet} = \{p_2\}$.

Definition 2.3 *The post-incidence and pre-incidence matrices of a net N are defined as integer matrices $[N]^+(p, t) = W(t, p)$ and $[N]^-(p, t) = W(p, t)$, respectively. The incidence matrix of N is defined as $[N] = [N]^+ - [N]^-$. The row of $[N]$ related to the place p is denoted by $[N](p, \cdot)$ and the column of $[N]$ related to the transition t is denoted by $[N](\cdot, t)$.*

The incidence matrix of a Petri net describes how places and transitions are linked in the net. Places are listed in the rows and transitions in the columns of an incidence matrix (vice versa for a transposed incidence matrix). A positive integer in an incidence matrix indicates an arc connecting from a transition to a place, while a negative one indicates an arc connecting from a place to a transition. In both cases, the absolute value of the integer denotes the weight of the arc. Note that zero in an incidence matrix indicates that no arc exists between a place and a transition or a self-loop between them, i.e., $W(t, p) = W(p, t)$.

Example 2.3 For the Petri net in Figure 2.1, its pre-incidence, post-incidence, and incidence matrices are respectively given as follows.

$$[N]^+ = \begin{bmatrix} 0 & 0 & 1 & 0 & 0 & 0 \\ 1 & 0 & 0 & 0 & 0 & 0 \\ 0 & 1 & 0 & 0 & 0 & 0 \\ 0 & 0 & 0 & 1 & 0 & 0 \\ 0 & 0 & 0 & 0 & 1 & 0 \\ 0 & 0 & 0 & 0 & 0 & 1 \\ 0 & 2 & 0 & 0 & 0 & 1 \\ 0 & 0 & 1 & 0 & 1 & 0 \end{bmatrix} \quad [N]^- = \begin{bmatrix} 1 & 0 & 0 & 0 & 0 & 0 \\ 0 & 1 & 0 & 0 & 0 & 0 \\ 0 & 0 & 1 & 0 & 0 & 0 \\ 0 & 0 & 0 & 0 & 1 & 0 \\ 0 & 0 & 0 & 0 & 0 & 1 \\ 0 & 0 & 0 & 1 & 0 & 0 \\ 2 & 0 & 0 & 0 & 1 & 0 \\ 0 & 1 & 0 & 1 & 0 & 0 \end{bmatrix}$$

$$[N] = \begin{bmatrix} -1 & 0 & 1 & 0 & 0 & 0 \\ 1 & -1 & 0 & 0 & 0 & 0 \\ 0 & 1 & -1 & 0 & 0 & 0 \\ 0 & 0 & 0 & 1 & -1 & 0 \\ 0 & 0 & 0 & 0 & 1 & -1 \\ 0 & 0 & 0 & -1 & 0 & 1 \\ -2 & 2 & 0 & 0 & -1 & 1 \\ 0 & -1 & 1 & -1 & 1 & 0 \end{bmatrix}$$

In the incidence matrix $[N]$, the entry -1 in the first row and the first column means that there is an arc with the weight one from p_1 to t_1. Similarly, the entry 2 in the seventh row and second column means that there exists an arc with the weight two from t_2 to p_7.

Definition 2.4 *For a net $N = (P, T, F, W)$, $M: P \to \mathbb{N}$ is called a marking that is a tokens' distribution in P. The number of tokens in a place p at a marking M is denoted by $M(p)$. A place p is said marked at M if $M(p) > 0$. A set of places $P' \subseteq P$ is said marked at M if $\exists p \in P'$, p is marked at M. M_0 is called an initial marking and (N, M_0) is called a net system of N.*

For conciseness, a compact multiset formalism $\sum_i M(p_i)p_i$ can be used to denote a marking M.

Example 2.4 For the Petri net shown in Figure 2.1, its initial marking $M_0 = (3, 0, 0, 0, 0, 3, 2, 1)^T$ can be expediently expressed by a multiset as $M_0 = 3p_1 + 3p_6 + 2p_7 + p_8$ since $M_0(p_1) = M_0(p_6) = 3$, $M_0(p_7) = 2$, $M_0(p_8) = 1$, and the rest places are not marked at M_0.

Definition 2.5 *At a marking M, if $\forall p \in {}^\bullet t$, $M(p) \geqslant W(p, t)$, i.e., $M \geqslant [N]^-(\cdot, t)$, we say that t is enabled at M, denoted as $M[t\rangle$. If t is not enabled at M, we write $\neg M[t\rangle$. If an enabled transition t fires at M, it generates another marking M' satisfying $\forall p \in P$, $M'(p) = M(p) - W(p, t) + W(t, p)$, i.e., $M' = M - [N]^-(\cdot, t) + [N]^+(\cdot, t) = M + [N](\cdot, t)$. Such a process is denoted by $M[t\rangle M'$. For a marking M, if firing a sequence of transitions σ finally generates a marking M', then M' is called reachable from M by firing σ, denoted as $M[\sigma\rangle M'$. The markings that are reachable from M is denoted by $R(N, M)$.*

Example 2.5 For the Petri net shown in Figure 2.1, t_1 and t_4 are enabled at the initial marking M_0, i.e., $M_0[t_1\rangle$ and $M_0[t_4\rangle$. If t_1 fires at M_0, a token is removed from p_1, two tokens are removed from p_7, and a token is added to p_2. Then, it generates a new marking $M_1 = 2p_1 + p_2 + 3p_6 + p_8$. This process is denoted as $M_0[t_1\rangle M_1$. If t_4 fires at M_0, another marking $M_2 = 3p_1 + p_4 + 2p_6 + 2p_7$ is generated, i.e., $M_0[t_4\rangle M_2$. Similarly, at M_1 and M_2, some enabled transitions exist and can fire to generate succeeding markings, which can be denoted as $M_1[t_2\rangle M_3$, $M_1[t_4\rangle M_4$, $M_2[t_1\rangle M_4$, and $M_2[t_5\rangle M_5$ with $M_3 = 2p_1 + p_3 + 3p_6 + 2p_7$, $M_4 = 2p_1 + p_2 + p_4 + 2p_6$, and $M_5 = 3p_1 + p_5 + 2p_6 + p_7 + p_8$. This process can continue until no new markings are generated. Then, all the markings constitute $R(N, M_0)$.

Definition 2.6 *Let σ be a finite sequence of transitions in a net $N = (P, T, F, W)$. The Parikh vector of σ is defined as $\vec{\sigma}: T \to \mathbb{N}$, which maps t in T to the number of occurrences of t in σ.*

Definition 2.7 *At any marking $M \in R(N, M_0)$, if firing a sequence of transitions σ finally generates a marking M', i.e., $M[\sigma\rangle M'$, we have*

$$M' = M + [N]\vec{\sigma} \tag{2.1}$$

which is called the state equation of (N, M_0).

Note that (2.1) is only a necessary condition for the reachability of a marking M' from M by firing σ, not a sufficient one, i.e., (2.1) does not imply $M[\sigma\rangle M'$.

Example 2.6 For the Petri net shown in Figure 2.1, we already know that $M_0[\sigma\rangle M_4$ with $\sigma = t_1 t_4$. So, we have

$$M_4 = M_0 + [N]\vec{\sigma} = \begin{bmatrix} 3 \\ 0 \\ 0 \\ 0 \\ 0 \\ 3 \\ 2 \\ 1 \end{bmatrix} + \begin{bmatrix} -1 & 0 & 1 & 0 & 0 & 0 \\ 1 & -1 & 0 & 0 & 0 & 0 \\ 0 & 1 & -1 & 0 & 0 & 0 \\ 0 & 0 & 0 & 1 & -1 & 0 \\ 0 & 0 & 0 & 0 & 1 & -1 \\ 0 & 0 & 0 & -1 & 0 & 1 \\ -2 & 2 & 0 & 0 & -1 & 1 \\ 0 & -1 & 1 & -1 & 1 & 0 \end{bmatrix} \begin{bmatrix} 1 \\ 0 \\ 0 \\ 0 \\ 1 \\ 0 \end{bmatrix} = \begin{bmatrix} 2 \\ 1 \\ 0 \\ 1 \\ 0 \\ 2 \\ 0 \\ 0 \end{bmatrix}.$$

Therefore, the generated marking is $M_4 = 2p_1 + p_2 + p_4 + 2p_6$.

Definition 2.8 *Suppose that $R(N, M_0)$ of a net system (N, M_0) is a finite set of markings. The reachability graph $G(N, M_0)$ of (N, M_0) is defined as a digraph (V, E) where $V = R(N, M_0)$ and $E = \{(M, t, M')|M, M' \in R(N, M_0) \wedge M[t\rangle M'\}$ are the sets of vertices and edges in the reachability graph, respectively. If $(M, t, M') \in E$, M is called an immediate predecessor of M' and M' is called an immediate successor of M in $G(N, M_0)$.*

The reachability graph helps us visualize the reachable markings and firing sequences of a Petri net and exactly reflects the complete behavior of the modeled system. Algorithm 2.1 constructs a reachability graph.

Algorithm 2.1

```
Input: A net system (N, M₀).
Output: The reachability graph G(N, M₀) of (N, M₀).
1. The root vertex of G(N, M₀) is the initial marking M₀. This ver-
tex is initially unlabeled;
2. While there exists any unlabeled vertex in G(N, M₀)
   {
        Select an unlabeled vertex M;
        for each transition t enabled at M
        {
            Compute the marking M' = M + [N](·, t);
            if (M' is not on G(N, M₀))
            {
                Add a new vertex M' to G(N, M₀);
                Add an edge labeled t from M to M' in G(N, M₀);
            }
        }
        Label vertex M;
   }
3. Return G(N, M₀).
```

Example 2.7   For the Petri net shown in Figure 2.1, its reachability graph is given in Figure 2.2 which has 10 vertices (markings) and 16 edges. Each edge connects two markings with a transition whose firing leads to the marking transfer between these two markings.

Note that reachability graphs are only for bounded Petri nets whose reachable markings are finite. For an unbounded Petri net which has infinite reachable markings, a finite approximation of its reachability graph, called a coverability graph, can be constructed [Hruz & Zhou, 2007]. Some recent studies about unbounded Petri nets can be found in (Fa *et al.*, 2019; Li *et al.* 2020).

2.2.2   Modeling Power of Petri Nets

Petri nets can efficiently model the typical characteristics in DEDSs, such as concurrency, synchronization, and mutual exclusion. The following are the elementary structures of Petri nets to model such activity characteristics.

2.2.2.1   Sequential Execution
In Figure 2.3a, transition t_2 can fire only after the firing of t_1. This imposes a precedence constraint that is typical in the execution of a DEDS.

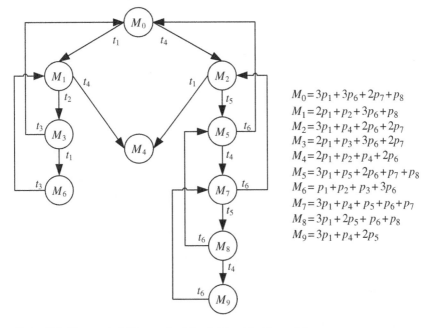

Figure 2.2 The reachability graph of the Petri net in Figure 2.1.

2.2.2.2 Concurrency (Parallelism)

The transitions t_1 and t_2 are concurrent in Figure 2.3b. A necessary condition for transitions to be concurrent is the existence of a forking transition that deposits a token in two or more post-places of the transition.

2.2.2.3 Synchronization

Usually, a part in a DEDS waits for resources or other parts to arrive (such as in an assembly line or an information fusion system). The resulting synchronization of activities can be depicted as in Figure 2.3c, in which t_1 is enabled only when both p_1 and p_2 have a token. The transition t_1 models a joining operation.

2.2.2.4 Conflict (choice)

In Figure 2.3d, t_1 and t_2 are in conflict. Both are enabled but the firing of one transition leads to the disabling of the other one. For example, such a situation may arise when a part has to choose among several machines in a manufacturing system.

2.2.2.5 Merging

When parts from different streams arrive for a service at the same machine, the resulting situation can be modeled as in Figure 2.3e.

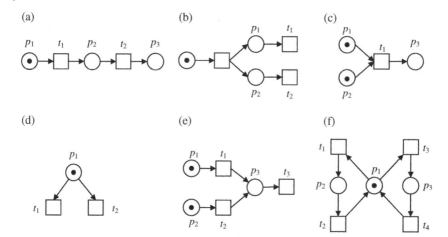

Figure 2.3 Elementary structures of Petri nets: (a) sequential execution; (b) concurrency; (c) synchronization; (d) conflict (choice); (e) merging; (f) mutual exclusion.

2.2.2.6 Mutual Exclusion

Two processes are mutually exclusive if they cannot be performed at the same time due to constraints on the usage of shared resources. Figure 2.3f shows such a case. For example, a robot may be shared by two machines for loading and unloading, but cannot perform both tasks simultaneously.

2.2.3 Behavioral Properties of Petri Nets

2.2.3.1 Boundedness and Safeness

Definition 2.9 *A place p in a net system (N, M_0) is said to be k-bounded if $\forall M \in R(N, M_0), M(p) \leq k$. (N, M_0) is k-bounded if all places are k-bounded. A 1-bounded net system is called a safe net system.*

Proposition 2.1 *A net system (N, M_0) is bounded if and only if it has a finite set of reachable markings.*

Example 2.8 For the Petri net shown in Figure 2.1, p_1 and p_6 are 3-bounded, p_5 and p_7 are 2-bounded, and the rest of the places are 1-bounded or safe. Thus, the net is 3-bounded and it has a finite number of reachable markings. However, it is not a safe net.

For RASs, the boundedness property may imply that no overflow occurs in a buffer, or can be used to limit the number of resources required by an operation.

2.2.3.2 Liveness and Deadlock

Definition 2.10 *For a net system* (N, M_0), *a transition* $t \in T$ *is said to be live at* M_0 *if* $\forall M \in R(N, M_0)$, $\exists M' \in R(N, M)$, $M'[t\rangle$. t *is dead at* $M \in R(N, M_0)$, *if* $\nexists M' \in R(N, M)$ *such that* $M'[t\rangle$. (N, M_0) *is live if* $\forall t \in T$, t *is live at* M_0. *A marking* $M \in R(N, M_0)$ *is called a deadlock marking if* $\nexists t \in T$, $M[t\rangle$. *A marking* $M \in R(N, M_0)$ *is called a livelock marking if* M *is not a deadlock marking,* $M_0 \notin R(N, M)$, *and* $\forall M' \in R(N, M)$, M' *is not a deadlock marking.* (N, M_0) *is deadlock-free (weakly live) if* $\forall M \in R(N, M_0)$, $\exists t \in T$, $M[t\rangle$.

Proposition 2.2 *Consider a bounded net system* (N, M_0) *and its reachability graph* $G(N, M_0)$. *A marking* $M \in R(N, M_0)$ *is a deadlock if and only if the vertex* M *in* $G(N, M_0)$ *has no output edge. A marking* $M \in R(N, M_0)$ *is a livelock if and only if the vertex* M *in* $G(N, M_0)$ *is not a deadlock and cannot reach* M_0 *or any deadlock.* (N, M_0) *is deadlock-free if and only if each vertex of* $G(N, M_0)$ *has at least one output edge.*

The reachability graph of a live or deadlock-free net system contains no dead markings. Importantly, a live net system is deadlock-free, but the converse is not true.

Example 2.9 The Petri net shown in Figure 2.1 is not live or deadlock-free since a dead marking M_4 exists in its reachability graph given in Figure 2.2.

2.2.3.3 Reversibility

Definition 2.11 *A net system* (N, M_0) *is said to be reversible if all of its reachable markings can return to the initial marking* M_0, *i.e.,* $\forall M \in R(N, M_0)$, $M_0 \in R(N, M)$.

Proposition 2.3 *A net system* (N, M_0) *is reversible if and only if* $G(N, M_0)$ *is a strongly connected graph.*

Example 2.10 For the reachability graph shown in Figure 2.2, M_4 cannot reach back to M_0. So, its corresponding Petri net as in Figure 2.1 is not reversible.

2.2.3.4 Conservativeness

Definition 2.12 *A net system* (N, M_0) *is said to be strictly conservative if at any reachable marking, the number of tokens in all places does not vary, i.e.,*

$$\forall M \in R(N, M_0), \quad \sum_{p_i \in P} M(p_i) = \sum_{p_i \in P} M_0(p_i). \tag{2.2}$$

This property implies that for any transition of the net the number of its input arcs equals that of its output arcs. A generalization of the strict conservativeness is as follows.

Definition 2.13 *A net system* (N, M_0) *is said to be conservative if there exists a vector X with* $|P|$ *positive integers such that*

$$\forall M \in R(N, M_0), \quad \sum_{p_i \in P} [X_i \cdot M(p_i)] = \sum_{p_i \in P} [X_i \cdot M_0(p_i)]. \tag{2.3}$$

Note that the strict conservativeness is a special case of the conservativeness in Definition 2.13 with $\forall i, X_i = 1$.

Proposition 2.4 *If a net system* (N, M_0) *is conservative then it is bounded.*

Note that conservativeness is only a sufficient condition for boundedness, but not a necessary condition.

Example 2.11 The Petri net shown in Figure 2.1 is neither strictly conservative nor conservative. It is bounded, however.

2.2.4 Subclasses of Petri Nets

This subsection presents different subclasses of Petri nets that satisfy particular structural conditions. The Venn diagram shown in Figure 2.4 summarizes the relations among these subclasses of Petri nets.

2.2.4.1 Ordinary Nets and Generalized Nets
Definition 2.14 *A Petri net* $N = (P, T, F, W)$ *is called an ordinary net if* $\forall (x, y) \in F, W(x, y) = 1$. *N is called a generalized net if* $\exists (x, y) \in F, W(x, y) > 1$.

Example 2.12 The Petri nets given in Figure 2.5 are all ordinary, while the net in Figure 2.1 is generalized since $W(p_7, t_1) = W(t_2, p_7) = 2$.

2.2.4.2 Pure Petri Nets
Definition 2.15 *A Petri net* $N = (P, T, F, W)$ *is said to be pure (self-loop free) if* $\forall p \in P, \forall t \in T, W(p, t) \cdot W(t, p) = 0$. *N is said to be restricted if it is both ordinary and pure.*

Example 2.13 The Petri nets given in Figure 2.5 are all restricted and the net shown in Figure 2.1 is pure but not restricted since it is not an ordinary net.

For a pure net, its net structure can be reconstructed according to its incidence matrix. However, for a non-pure net, its incidence matrix does not contain sufficient information to reconstruct the net since the matrix lacks self-loop information of a net.

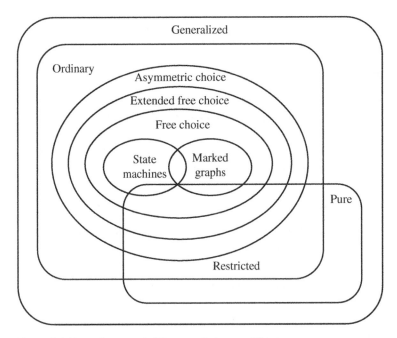

Figure 2.4 Venn diagram of different subclasses of Petri nets.

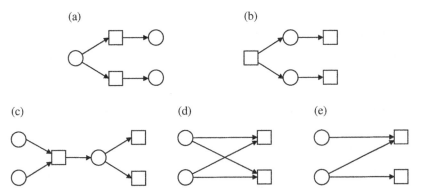

Figure 2.5 Subclasses of Petri nets. (a) State machine (b) marked graph (c) free-choice net (d) extended free-choice net (e) asymmetric choice net.

2.2.4.3 State Machines

Definition 2.16 *A state machine is an ordinary Petri net such that each transition has exactly one pre-place and one post-place, i.e., $\forall t \in T, |{}^\bullet t| = |t^\bullet| = 1$.*

The name of this subclass of Petri nets implies that these nets are structurally finite state automata.

Example 2.14 The net shown in Figure 2.5a is a state machine, while the rest nets in the figure and the net in Figure 2.1 are not.

2.2.4.4 Marked Graphs

Definition 2.17 *A marked graph is an ordinary Petri net such that each place has exactly one pre-transition and one post-transition, i.e., $\forall p \in P, |{}^{\bullet}p| = |p^{\bullet}| = 1$.*

Example 2.15 The net in Figure 2.5b is a marked graph, while the rest of the nets in the figure and the net in Figure 2.1 are not.

2.2.4.5 Free-choice Nets

Definition 2.18 *A free-choice net is an ordinary Petri net such that $\forall p_1, p_2 \in P$, if $p_1^{\bullet} \cap p_2^{\bullet} \neq \emptyset$, then $|p_1^{\bullet}| = |p_2^{\bullet}| = 1$.*

Importantly, both state machines and marked graphs are free-choice nets, but the converse is not true.

Example 2.16 The net shown in Figure 2.5c is a free-choice net, but it is neither a state machine nor a marked graph.

2.2.4.6 Extended Free-choice Nets

Definition 2.19 *An extended free-choice net is an ordinary Petri net such that $\forall p_1, p_2 \in P$, if $p_1^{\bullet} \cap p_2^{\bullet} \neq \emptyset$, then $p_1^{\bullet} = p_2^{\bullet}$.*

Example 2.17 The net shown in Figure 2.5d is an extended free-choice net.

2.2.4.7 Asymmetric Choice Nets

Definition 2.20 *An asymmetric choice net is an ordinary net such that $\forall p_1, p_2 \in P$, if $p_1^{\bullet} \cap p_2^{\bullet} \neq \emptyset$, then $p_1^{\bullet} \subseteq p_2^{\bullet}$ or $p_2^{\bullet} \subseteq p_1^{\bullet}$.*

Example 2.18 The net shown in Figure 2.5e is an asymmetric choice net.

2.2.5 Petri Nets for Resource Allocation Systems

Nowadays, many Petri nets have been proposed in the literature to model RASs. The relationships among different classes of Petri nets for RASs are shown in Figure 2.6.

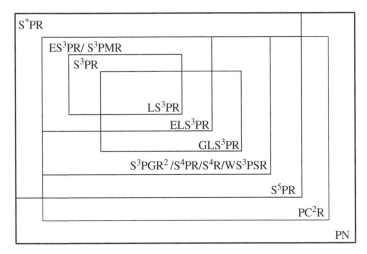

Figure 2.6 Relations among different classes of PNs for RASs.

2.2.5.1 PC²R

PC^{2}R denotes processes competing for conservative resources. PC^{2}R nets are usually used to model and analyze concurrent programming in software engineering systems. To define PC^{2}R, we first introduce the definition of iterative state machines.

Definition 2.21 *An iterative state machine $N = (P, T, F, W)$ is a strongly connected state machine such that i) P is classified into three subsets: $\{p_k\}$, P', and P''; ii) $P' \neq \emptyset$; iii) the subnet generated by $\{p_k\} \cup P'$ and $\bullet P' \cup P'^\bullet$ is a strongly connected state machine in which every cycle contains p_k; and iv) if $P'' \neq \emptyset$, the subnet generated by $\{p_k\} \cup P''$ and $\bullet P'' \cup P''^\bullet$ is also an iterative state machine.*

Definition 2.22 *A PC^{2}R net $N = (P_0 \cup P_R \cup P_A, T, F, W)$ is defined as the composition of a set of subnets $N^x = (\{p_0^x\} \cup P_R^x \cup P_A^x, T^x, F^x, W^x)$ that share some common places, i.e., $N = \bigcup_{x \in \mathbb{N}_J} N^x$ where $\mathbb{N}_J = \{i \in \mathbb{Z}^+ | N^i$ is the ith processing subnet of $N\}$ and the following statements are satisfied.*

1) *p_0^x is called an idle place of N^x. Elements in P_A^x and P_R^x are called activity (operation) and resource places of N^x, respectively.*
2) *$P_A^x \neq \emptyset$; $P_R^x \neq \emptyset$; $p_0^x \notin P_A^x$; $(\{p_0^x\} \cup P_A^x) \cap P_R^x = \emptyset$.*
3) *$W = W_A \cup W_R$ where $W_A: ((P_A \cup P_0) \times T) \cup (T \times (P_A \cup P_0)) \to \{0,1\}$ such that $\forall x, y \in \mathbb{N}_J$ with $x \neq y$, $((P_A^y \cup \{p_0^y\}) \times T^x) \cup (T^x \times (P_A^y \cup \{p_0^y\})) \to \{0\}$ and $W_R: (P_R \times T) \cup (T \times P_R) \to \mathbb{N}$.*
4) *N^x and $N^y (x \neq y)$ are composable by sharing a set of common places that belong to P_R.*

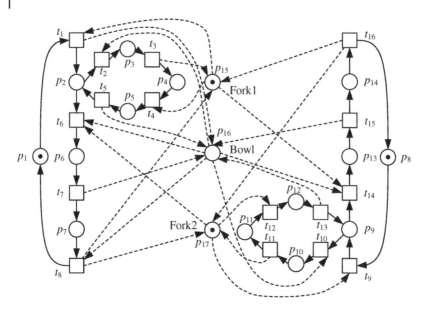

Figure 2.7 A PC²R net for a postmodern dining philosophers problem.

5) $\forall r \in P_R$, there exists a unique minimal P-semiflow I_r such that $\|I_r\| \cap P_R = \{r\}$, $\|I_r\| \cap (P_0 \cup P_A) \neq \emptyset$, and $I_r(r) = 1$. In addition, $P_A \subseteq \bigcup_{r \in P_R}(\|I_r\| \setminus \{r\})$.

6) If the resource places of P_R^x and their related arcs are removed from N^x, the resultant net $N^{x\prime}$ is an iterative state machine.

A PC²R net is also a pure and generalized net. More importantly, its processing subnets may contain internal iterations that can represent recirculating circuits in manufacturing systems or nested loops in software design. In addition, its initial marking may have some resource units that have already been allocated.

Example 2.19 The net shown in Figure 2.7 is a PC²R net that models an improved postmodern dining philosophers problem. It has two nested iterative state machines in its processing subnets and the allocation of resource p_{17} is no longer first-acquire-later-release, i.e., the resource has already been allocated at the initial marking.

2.2.5.2 S*PR
S*PR is used to model sequential RAS with alternative routes and alternative requests for conjunctive-type resources.

Definition 2.23 An S*PR net $N = (P_0 \cup P_R \cup P_A, T, F, W)$ is defined as the composition of a set of subnets $N^x = (\{p_0^x\} \cup P_R^x \cup P_A^x, T^x, F^x, W^x)$ sharing some common places, i.e., $N = \bigcup_{x \in \mathbb{N}_J} N^x$, and the following statements are true.

1) p_0^x is called an idle place of N^x. Elements in P_A^x and P_R^x are called activity (operation) and resource places of N^x, respectively.

2) $P_A^x \neq \emptyset$; $P_R^x \neq \emptyset$; $p_0^x \notin P_A^x$; $(\{p_0^x\} \cup P_A^x) \cap P_R^x = \emptyset$.

3) $W = W_A \cup W_R$ where $W_A: ((P_A \cup P_0) \times T) \cup (T \times (P_A \cup P_0)) \rightarrow \{0, 1\}$ such that $\forall x, y \in \mathbb{N}_J$ with $x \neq y$, $((P_A^y \cup \{p_0^y\}) \times T^x) \cup (T^x \times (P_A^y \cup \{p_0^y\})) \rightarrow \{0\}$ and $W_R: (P_R \times T) \cup (T \times P_R) \rightarrow \mathbb{N}$.

4) N^x and $N^y (x \neq y)$ are composable by only sharing a set of common places that are resource places.

5) $\forall r \in P_R$, there exists a unique minimal P-semiflow I_r such that $\|I_r\| \cap P_R = \{r\}$, $\|I_r\| \cap P_0 = \emptyset$, $\|I_r\| \cap P_A \neq \emptyset$, and $I_r(r) = 1$. In addition, $P_A = \bigcup_{r \in P_R}(\|I_r\| \setminus \{r\})$.

6) If the resource places in P_R^x and their related arcs are removed from N^x, the resultant net $N^{x\prime}$ is a strongly connected state machine.

An S*PR net is also pure and generalized. Compared with PC^{2}R, the processing subnets of S*PR can be any kind of strongly connected state machines including iterative state machines and its initial marking has no resources that have already been allocated.

2.2.5.3 S^{5}PR

S^{5}PR is a subclass of PC^{2}R and its initial marking has no resources that have already been allocated.

Definition 2.24 *An S^{5}PR net is a PC^{2}R net such that $\forall r \in P_R$, $\|I_r\| \cap P_0 = \emptyset$.*

It is important to note that S^{5}PR is similar to S*PR except that the processing subnets of S^{5}PR are iterative state machines that belong to a kind of strongly connected state machines while the processing subnets of S*PR can be any kind of strongly connected state machines. Thus, S^{5}PR is a subclass of S*PR.

2.2.5.4 S^{4}PR, S^{4}R, S^{3}PGR^2 and WS^{3}PSR

S^{4}PR (S^{4}R) is a pure and generalized net and represents a system of sequential systems with shared resources. S^{3}PGR^2 denotes a system of simple sequential processes with general resource requirements and WS^{3}PSR denotes a weighted system of simple sequential processes with several resources. These four classes of Petri nets are the same by definition.

Definition 2.25 *An S^{4}PR / S^{4}R / S^{3}PGR^2 / WS^{3}PSR net is an S^{5}PR net such that $\forall x \in \mathbb{N}_J$, the subnet generated by $\{p_0^x\} \cup P_A^x$ and T^x is a strongly connected state machine in which every cycle contains the idle place p_0^x.*

That is to say, S^{5}PR is a superclass of S^{4}PR / S^{4}R / S^{3}PGR^2 / WS^{3}PSR since it allows nested loops in which some cycles may not contain an idle place.

2.2.5.5 S³PR

S³PR denotes a system of simple sequential processes with resources.

Definition 2.26 *An S³PR net is an S⁴PR net such that $\forall p \in P_A$, $^{\bullet\bullet}p \cap P_R = p^{\bullet\bullet} \cap P_R$, $|^{\bullet\bullet}p \cap P_R| = 1$, and $W_R: (P_R \times T) \cup (T \times P_R) \rightarrow \{0,1\}$.*

S³PR is ordinary and each of its operation places requires at most a single resource unit.

Example 2.20 The net shown in Figure 2.8 is an S³PR net that models a flexible manufacturing system (Zhou *et al.*, 1992).

2.2.5.6 ES³PR and S³PMR

ES³PR denotes an extended S³PR while S³PMR represents a system of simple sequential processes with multiple resources. They are equivalent and contain of S³PR.

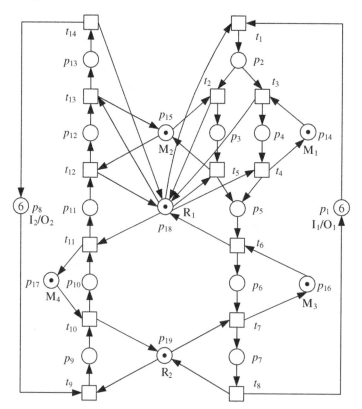

Figure 2.8 An S³PR net for an FMS.

Definition 2.27 *ES³PR / S³PMR is an extension to S³PR without the restriction that* $\forall p \in P_A$, $|^{\bullet\bullet}p \cap P_R| = 1$.

That is to say, an operation of ES³PR/S³PMR may need several resource units, each of which belongs to a different resource type. Note that ES³PR/S³PMR is also ordinary.

2.2.5.7 LS³PR

LS³PR denotes a system of linear sequential processes with resources.

Definition 2.28 *An LS³PR net is an S³PR net such that* $\forall p \in P_A$, $|p^{\bullet}| = 1$.

It means that LS³PR is a subclass of S³PR where all activity places have no choice (conflict). Note that idle places of LS³PR may have choices.

Example 2.21 Figure 2.9 gives some examples of processing subnets that show the difference between LS³PR and S³PR. Note that resource places and their arcs are not shown in this figure for conciseness.

2.2.5.8 ELS³PR

ELS³PR represents an extended system of linear sequential processes with resources.

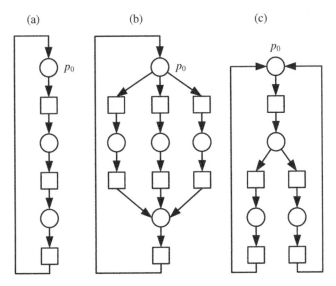

(a)   (b)   (c)

Figure 2.9 Some processing subnets. (a) and (b) are subnets of LS³PR; (c) is a subnet of S³PR (not LS³PR).

Definition 2.29 *ELS^3PR is an extension of LS^3PR without the restriction that $\forall p \in P_A$, $|^{\bullet\bullet}p \cap P_R| = 1$.*

Thus, ELS^3PR properly includes LS^3PR and it is a subclass of ES^3PR.

2.2.5.9 GLS³PR

GLS^3PR denotes a system of generalized linear simple sequential processes with resources.

Definition 2.30 *GLS^3PR is an extension of ELS^3PR such that $W_R\colon (P_R \times T) \cup (T \times P_R) \to \mathbb{N}$.*

GLS^3PR is generalized and it properly includes ELS^3PR.

Example 2.22 The net shown in Figure 2.1 is a GLS^3PR net.

There also exist some other classes of Petri nets for RASs such as PPN and G-systems. PPN, which represents a production Petri net, is similar to LS^3PR except that after removing resources, the processing subnets of PPN are acyclic and have no idle places. G-systems, on the other hand, are generalized nets with source and sink places and the processing subnets are not limited to state machines and allow assembly and disassembly operations.

2.2.6 Structural Analysis

Structural analysis methods are often used to control Petri net models via specific net structures such as siphons, traps, place invariants, and resource-transition circuits. Based on these structures, behavioral properties (e.g. liveness and deadlocks) are usually analyzed efficiently. This subsection presents the definitions and features of classical Petri net structures used in structural analysis.

Definition 2.31 *Given a Petri net $N = (P, T, F, W)$, a place invariant $I\colon P \to \mathbb{Z}$ is defined as a $|P|$-dimensional vector such that $I \neq \mathbf{0}$ and $I^T[N] = \mathbf{0}^T$; a transition invariant $J\colon T \to \mathbb{Z}$ is defined as a $|T|$-dimensional vector such that $J \neq \mathbf{0}$ and $[N]J = \mathbf{0}$.*

Proposition 2.5 *Given a net system (N, M_0), let I be a place invariant of the net. We have*

$$\forall M \in R(N, M_0), I^T M = I^T M_0. \tag{2.4}$$

The elements of a place invariant can be interpreted as weights of their corresponding places. So, Proposition 2.5 indicates that the total number of weighted tokens in the places of a place invariant remains constant at any reachable marking from the initial marking.

Proposition 2.6 *Given a net system (N, M_0), let M and M' be two reachable markings in $R(N, M_0)$ and σ be a firing sequence such that $M[\sigma\rangle M'$. Then $\vec{\sigma}$ is a transition invariant if and only if $M = M'$.*

The elements of a transition invariant can be interpreted as the firing numbers of their corresponding transitions. So, Proposition 2.6 indicates that firing all transitions as many times as their firing numbers in a transition invariant leads the model to the same marking. Note that all place invariants (or transition invariants) form a linear space that is closed under addition, scalar multiplication by integers, and division by the greatest common divisor of all elements in them.

Definition 2.32 *A place invariant I is called a P-semiflow if each element of I is non-negative. $||I|| = \{p|I(p) \neq 0\}$ is called the support of I. If all the elements of I are coprime and $||I||$ is not a superset of the support of any other place invariant, then I is called a minimal place invariant.*

Example 2.23 For the Petri net shown in Figure 2.1, there exist at least four place invariants $I_1 = (1\ 1\ 1\ 0\ 0\ 0\ 0\ 0)^T$, $I_2 = (0\ 0\ 0\ 1\ 1\ 1\ 0\ 0)^T$, $I_3 = (0\ 2\ 0\ 0\ 1\ 0\ 1\ 0)^T$, and $I_4 = (0\ 0\ 1\ 1\ 0\ 0\ 0\ 1)^T$ and two transition invariants $J_1 = (1\ 1\ 1\ 0\ 0\ 0)^T$ and $J_2 = (0\ 0\ 0\ 1\ 1\ 1)^T$. We have $||I_1|| = \{p_1, p_2, p_3\}$, $||I_2|| = \{p_4, p_5, p_6\}$, $||I_3|| = \{p_2, p_5, p_7\}$, and $||I_4|| = \{p_3, p_4, p_8\}$ and all of them are P-semiflows and minimal place invariants.

Note that we can use a compact multiset formalism $\sum_i M(p_i)p_i$ to denote a place invariant, for example, $I_1 = p_1 + p_2 + p_3$ and $I_3 = 2p_2 + p_5 + p_7$.

Definition 2.33 *Given a Petri net $N = (P, T, F, W)$, a siphon is a set of places $P' \subseteq P$ such that ${}^{\bullet}P' \subseteq P'^{\bullet}$. A trap is a set of places $P' \subseteq P$ such that $P'^{\bullet} \subseteq {}^{\bullet}P'$. A siphon (trap) is minimal if it is not the superset of any other siphon (trap). A minimal siphon is said to be strict if ${}^{\bullet}P' \subsetneq P'^{\bullet}$.*

Example 2.24 For the Petri net shown in Figure 2.1, $P' = \{p_3, p_5, p_7, p_8\}$ is a siphon since ${}^{\bullet}P' \subseteq P'^{\bullet}$ with ${}^{\bullet}P' = \{t_2, t_3, t_5, t_6\}$ and $P'^{\bullet} = \{t_1, t_2, t_3, t_4, t_5, t_6\}$. In addition, P' is a minimal siphon since there exists no other siphon that is a proper subset of P'. P' is also a strict siphon since ${}^{\bullet}P' \subsetneq P'^{\bullet}$.

Proposition 2.7 *Given a net system (N, M_0), let M be a reachable marking $M \in R(N, M_0)$ and P' be a siphon. $\forall p \in P'$, if $M(p) = 0$, then $\forall M' \in R(N, M), M'(p) = 0$.*

Proposition 2.7 means that once a siphon becomes empty at a marking, it remains empty at all the markings reachable from the marking.

Proposition 2.8 *Given a net system* (N, M_0), *let M be a reachable marking* $M \in R(N, M_0)$ *and P' be a trap.* $\forall p \in P'$, *if* $M(p) > 0$, *then* $\forall M' \in R(N, M)$, $M'(p) > 0$.

Proposition 2.8 indicates that once a trap becomes marked at a marking, it remains marked at all the markings reachable from the marking.

Proposition 2.9 *Let* (N, M_0) *be an ordinary net and* Π *be the set of its siphons. The net is deadlock-free if* $\forall M \in R(N, M_0)$, $\forall P' \in \Pi$, $\exists p \in P'$, $M(p) > 0$.

2.2.7 Reachability Graph Analysis

The reachability graph of a Petri net reflects the underlying system's complete behavior, i.e., it exactly represents the possible states of the system. Thus, a reachability graph is important for the optimal control and scheduling of the underlying system. This section presents the reachability graph analysis methods for the supervisory control and system scheduling of RASs.

2.2.7.1 Supervisory Control

Consider a Petri net N for RASs. Its reachability graph $G(N, M_0)$ consists of two disjointed parts: a live zone (LZ) and a deadlock zone (DZ). The markings in LZ are legal markings that can reach the initial marking M_0 from at least one of their successors. The set of legal markings is defined as:

$$\mathcal{M}_L = \{M | M \in R(N, M_0) \wedge M_0 \in R(N, M)\}. \tag{2.5}$$

The rest of the graph is DZ that consists of deadlock markings, livelock markings, and bad markings that inevitably lead to deadlocks and livelocks. A first-met bad marking (FBM) is a marking in DZ and it is an immediate successor of a marking in LZ. The set of FBMs can be defined as:

$$\mathcal{M}_F = \{M \in DZ | \exists M' \in LZ, \ t \in T, \text{s.t. } M'[t\rangle M\}. \tag{2.6}$$

Consider a resource allocation system that has three shared resources R_1–R_3 (such as machines and robots), two loading buffers I_1 and I_2, and two unloading buffers O_1 and O_2. Suppose that a resource only processes one part at a time. Two types of parts, P_1 and P_2, are to be processed in the system. The production sequences are:

$$P_1 : \ I_1 \rightarrow R_1 \rightarrow R_2 \ (\text{or } R_3) \rightarrow R_3 \rightarrow O_1$$
$$P_2 : \ I_2 \rightarrow R_3 \rightarrow R_2 \rightarrow R_1 \rightarrow O_2$$

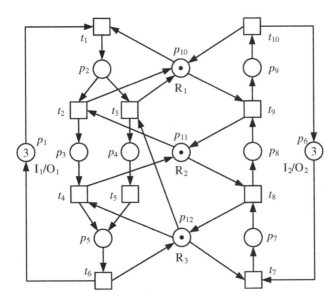

Figure 2.10 Petri net model of a resource allocation system.

Figure 2.10 shows its Petri net model that has 12 places and 10 transitions. Its reachability graph is given in Figure 2.11. For this reachability graph, we have that M_9, M_{11}, and M_{22} constitute DZ and the rest markings belong to LZ. Note that all the markings in DZ are FBMs for this system. In addition, there are two deadlock markings $M_9 = p_1 + p_2 + p_3 + 2p_6 + p_7$ and $M_{22} = 2p_1 + p_2 + p_6 + p_7 + p_8$ since they have no successors. If all the FBMs are forbidden and all legal markings are permitted by an added supervisor, the resulting controlled net is kept running in LZ and never enters DZ, i.e., the controlled net with the supervisor is deadlock-free. Furthermore, the reachability graph of the controlled net becomes a strongly connected component contains all transitions. Therefore, the controlled net is also live. Since all legal markings survive in the controlled net, the added supervisor is maximally permissive (optimal) in behavior. To forbid FBMs and permit legal markings, place invariants need to be constructed to design a supervisor with control places and related arcs (Uzam & Zhou, 2006).

2.2.7.2 System Scheduling

Once the Petri net model of a resource allocation system is constructed, the evolution of the system can be described by the marking transfer in the reachability graph, that is, all possible behaviors of the system can be completely tracked by the reachability graph. Thus, given an initial marking and a goal marking of the net, an optimal system schedule can be obtained by generating the reachability graph and finding an optimal path from the initial marking to the goal marking. The

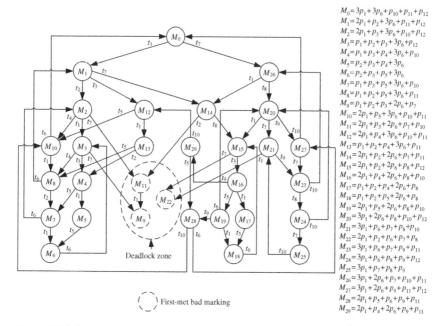

$M_0=3p_1+3p_6+p_{10}+p_{11}+p_{12}$
$M_1=2p_1+p_2+3p_6+p_{11}+p_{12}$
$M_2=2p_1+p_3+3p_6+p_{10}+p_{12}$
$M_3=p_1+p_2+p_3+3p_6+p_{12}$
$M_4=p_1+p_3+p_4+3p_6+p_{10}$
$M_5=p_2+p_3+p_4+3p_6$
$M_6=p_2+p_3+p_5+3p_6$
$M_7=p_1+p_3+p_5+3p_6+p_{10}$
$M_8=p_1+p_2+p_5+3p_6+p_{11}$
$M_9=p_1+p_2+p_3+2p_6+p_7$
$M_{10}=2p_1+p_5+3p_6+p_{10}+p_{11}$
$M_{11}=2p_1+p_3+2p_6+p_7+p_{10}$
$M_{12}=2p_1+p_4+3p_6+p_{10}+p_{11}$
$M_{13}=p_1+p_2+p_4+3p_6+p_{11}$
$M_{14}=2p_1+p_2+2p_6+p_7+p_{11}$
$M_{15}=2p_1+p_2+2p_6+p_8+p_{12}$
$M_{16}=2p_1+p_4+2p_6+p_8+p_{10}$
$M_{17}=p_1+p_2+p_4+2p_6+p_8$
$M_{18}=p_1+p_2+p_5+2p_6+p_8$
$M_{19}=2p_1+p_5+2p_6+p_8+p_{10}$
$M_{20}=3p_1+2p_6+p_8+p_{10}+p_{12}$
$M_{21}=3p_1+p_6+p_7+p_8+p_{10}$
$M_{22}=2p_1+p_2+p_6+p_7+p_8$
$M_{23}=3p_1+p_6+p_7+p_9+p_{11}$
$M_{24}=3p_1+p_6+p_8+p_9+p_{12}$
$M_{25}=3p_1+p_7+p_8+p_9$
$M_{26}=3p_1+2p_6+p_7+p_{10}+p_{11}$
$M_{27}=3p_1+2p_6+p_9+p_{11}+p_{12}$
$M_{28}=2p_1+p_5+p_6+p_9+p_{11}$
$M_{29}=2p_1+p_4+2p_6+p_9+p_{11}$

Figure 2.11 Reachability graph of the Petri net in Figure 2.10.

obtained path is a firing sequence of transitions in the Petri net. However, even for a simple Petri net model, its reachability graph may be very large. To handle it, Artificial Intelligence (AI) based heuristic search algorithms can be employed to find such a path without exploring the whole reachability graph. For example, an often used heuristic search algorithm is the A* search that generates the necessary portion of the reachability graph to find an optimal path with an admissible heuristic function. In this way, an optimal schedule is possible to be obtained in a reasonable amount of time if such a schedule exists.

2.2.8 Petri Net Analysis Tools

There are many software tools that can model and simulate Petri nets. However, most of the tools are graphical editors that implement token game animation. Only a few tools, such as INA (Starke, 2019), CPN tools (Beaudouin-Lafon *et al.*, 2001), JFern (2018), Petruchio (Meyer & Strazny, 2010), and TINA (Berthomieu *et al.*, 2004), have all the following features: generating net structures and state spaces, dealing with timed Petri nets, and doing performance analysis. Among them, INA is a very popular tool for the analysis of Petri nets. Besides the aforementioned abilities, INA can deal with place invariants, transition invariants, and net reduction, and perform structural analysis and advanced performance analysis.

To analyze a Petri net in INA, the net should be given as an input file. In the input file of the net, each place of the net is described by a line that starts with the ID number of the place, followed by a blank and the number of tokens in the place at the initial marking. Next, each pre-transition of the place is written as *<transition ID number>* : *<arc weight>* in which : *<arc weight>* if omitted if the arc weight equals one. Any pair of adjacent pre-transitions of the place are separated by a blank. The list of pre-transitions is empty if the place has no pre-transitions. If the place has post-transitions, a similar list of post-transitions preceded by <, > has to be entered. Finally, each line is ended with *<cr>*. For example, the input file of INA for the Petri net shown in Figure 2.1 is given as follows:

```
p M pre, post
1 3 3,1
2 0 1,2
3 0 2,3
4 0 4,5
5 0 5,6
6 3 6,4
7 2 2:2 6,1:2 5
8 1 3 5,2 4
@
```

By using INA, the resulting analysis of the net is given below:

```
Computation of the reachability graph
States generated: 10
Arcs generated: 16
Number of dead states found: 1
The net has dead reachable states.
The net is not live.
The net is not reversible (resettable).
The net has no dead transitions at the initial mark-
```
ing.
```
The net is not live, if dead transitions are ignored.
The net is bounded.
```

The reachability graph of the net can also be computed by INA as:

```
State nr.     1
P.nr: 1 2 3 4 5 6 7 8
toks: 3 0 0 0 0 3 2 1
==t1=> s2
==t4=> s6
State nr.     2
```

```
P.nr: 1 2 3 4 5 6 7 8
toks: 2 1 0 0 0 3 0 1
==t2=> s3
==t4=> s5
State nr.    3
P.nr: 1 2 3 4 5 6 7 8
toks: 3 0 0 1 0 2 2 0
==t1=> s5
==t5=> s7
State nr.    4
P.nr: 1 2 3 4 5 6 7 8
toks: 2 0 1 0 0 3 2 0
==t1=> s4
==t3=> s1
State nr.    5
P.nr: 1 2 3 4 5 6 7 8
toks: 2 1 0 1 0 2 0 0
dead state
State nr.    6
P.nr: 1 2 3 4 5 6 7 8
toks: 3 0 0 0 1 2 1 1
==t4=> s8
==t6=> s1
State nr.    7
P.nr: 1 2 3 4 5 6 7 8
toks: 1 1 1 0 0 3 0 0
==t3=> s2
State nr.    8
P.nr: 1 2 3 4 5 6 7 8
toks: 3 0 0 1 1 1 1 0
==t5=> s9
==t6=> s6
State nr.    9
P.nr: 1 2 3 4 5 6 7 8
toks: 3 0 0 0 2 1 0 1
==t4=> s10
==t6=> s7
State nr.    10
P.nr: 1 2 3 4 5 6 7 8
toks: 3 0 0 1 2 0 0 0
==t6=> s8
```

The reachability graph is depicted as in Figure 2.2 where M_4 is both a deadlock marking and FBM. Note that the subscripts of states (markings) in the output file of INA start from 1. So, state 1 refers to M_0 in Figure 2.2, state 2 refers to M_1, and so on.

2.3 Informed Heuristic Search

2.3.1 Basic Concepts of Heuristic A* Search

Given a Petri net N of a resource allocation system, its reachability graph $G(N, M_0)$ is an OR graph. In the search spaces of OR graphs, several informed heuristic search strategies can be used, such as best-first (BF), BF*, Z, Z*, and A*, all of which use heuristic information to decide which node to expand next (Pearl, 1984; Russell & Norvig, 2010). Among them, A* is the most popular algorithm. Algorithm 2.2 gives its explicit description, followed by some explanations. The detailed properties of A* are given in Section 2.3.2.

Algorithm 2.2

```
A* Algorithm:
    1)  Put the start node n₀ on the list OPEN.
    2)  If OPEN is empty, terminate with failure.
    3)  Remove the first node n from OPEN and put n on the list CLOSED.
    4)  If n equals a goal node n_G, construct the path by tracing back
the pointer from n to n₀ and terminate.
    5)  Expand n, generate all its successors, and set their point-
ers back to n.
    6)  For every successor n´ of n, compute g(n´):
        a)          If n´ is already on OPEN, direct its pointer along the
path yielding the smallest g(n´).
        b)          If n´ is already on CLOSED, direct its pointer along the
path yielding the smallest g(n´). if n´ requires pointer redirection, move
n´ to OPEN.
        c)          If n´ is not on OPEN and CLOSED, calculate h(n´) and f(n´) =
g(n´) + h(n´), and put n´ on OPEN.
    7)  Reorder OPEN in the non-decreasing magnitude of f.
    8)  Go to Step 2).
```

The A* algorithm finds a path in a graph as follows. It uses an evaluation function for a node n as $f(n) = g(n) + h(n)$ in which $f(n)$ is an estimate of the lowest cost $f^*(n)$ from n_0 to n_G among all paths going through the current node n, $g(n)$ is the current lowest cost obtained from n_0 to n, and $h(n)$ is called a heuristic function that is an estimate of the lowest cost $h^*(n)$ from n to n_G among all paths. In the A* algorithm, the list OPEN contains nodes that are generated but not expanded and the list CLOSED contains nodes that have been expanded. At each iteration, a node with the smallest f-value in OPEN is selected for expansion to generate successor nodes. If more than one node has the smallest f-value in OPEN, a tie-breaking rule is applied to choose one of the candidates for expansion. It iteratively expands the search space until a given goal node is selected for expansion. Then, a schedule is constructed by tracing the pointers from the goal node back to the initial one. The order of the activities, i.e., the system schedule, is thus obtained.

2.3.2 Properties of the A* Search

2.3.2.1 Completeness

Definition 2.34 *An algorithm is said to be complete if it can terminate with a result whenever a solution exists.*

Proposition 2.10 *A^* is complete on both finite and infinite graphs.*

2.3.2.2 Admissible Heuristics

Definition 2.35 *In an A^* algorithm, a heuristic function h is said to be admissible if for any reachable node n, $h(n) \leq h^*(n)$.*

Proposition 2.11 *If an A^* algorithm uses an admissible heuristic function, its obtained result is guaranteed to be optimal.*

2.3.2.3 Monotone (Consistent) Heuristics

Definition 2.36 *In an A^* algorithm, a heuristic function h is said to be monotone or consistent if for any edge (n, n') in the search space, $h(n) \leq c(n, n') + h(n')$ and $h(n_G) = 0$ where $c(n, n')$ denotes the cost of the edge (n, n').*

Proposition 2.12 *A monotone heuristic function is admissible.*

Proposition 2.13 *An A^* algorithm with a monotone heuristic function finds optimal paths to all expanded nodes, i.e., $\forall n \in CLOSED$, $g(n) = g^*(n)$ where $g^*(n)$ denotes the lowest cost among all paths going from n_0 to n.*

Proposition 2.14 *For an A^* algorithm with a monotone heuristic function, the f-values of the sequence of expanded nodes are non-decreasing.*

2.3.2.4 More Informed Heuristics

To efficiently search in a graph with an A* algorithm, different heuristics may be designed and tested. The used heuristic function h is important to its efficiency since its pruning power depends on the accuracy of h.

Definition 2.37 *Given two heuristic functions h_1 and h_2, if each node expanded by an A^* algorithm with h_1 is also expanded by an A^* algorithm with h_2, then h_1 is said to be more informed than h_2.*

Note that the phrase "more informed than" is often used interchangeably with "more efficient than" in literature.

Proposition 2.15 *Given two heuristic functions h_1 and h_2, h_1 is more informed than heuristic h_2 if both are admissible and for each nongoal node n, $h_1(n) > h_2(n)$.*

The use of a more informed heuristic in A^* implies more pruning power, that is to say, the higher the h-value, the closer h is to h^* (since h is admissible), which yields a more efficient search. If h estimates the cost precisely, i.e., $h = h^*$, then the A^* algorithm with h^* only expands nodes lying along optimal paths.

2.4 Bibliographical Notes

Additional details on Petri nets can be found in a survey paper of Murata (1989) and the books (Hrúz & Zhou, 2007; Li & Zhou, 2008; WU & Zhou, 2010; Zhou & Venkatesh, 1999). Interesting results on the efficient computation of siphons can be found in Cordone *et al.* (2005), Piroddi *et al.* (2008), You *et al.*, (2017b). For the applications of different classes of Petri nets for RASs, please refer to PC^2R (López-Grao & Colom, 2012), S^*PR (Ezpeleta *et al.*, 2002), S^5PR (López-Grao & Colom, 2006), S^4PR (Tricas *et al.*, 1999), S^4R (Barkaoui & Abdallah, 1996), S^3PGR^2 (Park & Reveliotis, 2001), WS^3PSR (Tricas *et al.*, 1995), S^3PR (Ezpeleta *et al.*, 1995), ES^3PR (Tricas *et al.*, 1998), S^3PMR (Huang *et al.*, 2006), LS^3PR (Ezpeleta *et al.*, 1998), ELS^3PR (Wang *et al.*, 2011), GLS^3PR (Hou *et al.*, 2014), and G-systems (Li & Zhao, 2008; Zouari & Barkaoui, 2003). Li *et al.* (2008b, 2012) and Du *et al.* (2020) give surveys and comparisons of deadlock control methods based on Petri nets. The concepts of LZ, DZ, and FBMs are developed in Uzam and Zhou (2007) and are then used in the literature to obtain an optimal supervisor synthesis for RASs. For more intelligent search strategies within graphs, please refer to the books of Pearl (1984) and Edelkamp and Schroedl (2012). The summary of tools for Petri nets' modeling and analysis can be found in Uni-Hamburg (2019). A detailed manual of INA can be seen in Starke (2019).

Part II

Supervisory Control

3

Behaviorally Maximal and Structurally Minimal Supervisor

This chapter presents applications of the theory of regions in synthesizing optimal Petri net supervisors with the fewest monitors for resource allocation systems (RASs). Petri nets that can be used by the method are denoted as DP-nets in this book where DP stands for deadlock prevention. The presented method first calculates the sets of legal markings and first-met bad markings (FBMs) by analyzing the reachability graph of a DP-net. Next, the sets of markings are reduced to two small ones by using a vector covering method. Then, some place invariants, each of which is associated with a monitor that contains a control place and several arcs, are designed to permit all legal markings and forbid all FBMs via an integer linear program that minimizes the number of added monitors in the supervisor. It guarantees that the added supervisor is structurally minimal and the controlled net is optimal, i.e., behaviorally maximal. To demonstrate the usage of the method, an illustrative example is given in the chapter.

3.1 Introduction

Behavioral permissiveness and structural complexity are two important criteria used to evaluate a Petri net supervisor designed for a given resource allocation system. A maximally permissive (optimal) supervisor implies that it allows the reachability of all legal markings and forbids all illegal markings including deadlock markings, livelock markings, and bad markings that inevitably evolve into deadlocks and livelocks. Such a supervisor can lead to the high utilization of system resources. On the other hand, a structure-minimal supervisor can decrease the hardware and software costs in control validation and implementation stages of the underlying system.

To prevent all illegal states and permit all legal behaviors of RASs, Uzam (2002) adopts the theory of regions to derive optimal Petri net supervisors if such

Supervisory Control and Scheduling of Resource Allocation Systems: Reachability Graph Perspective, First Edition. Bo Huang and MengChu Zhou.
© 2020 The Institute of Electrical and Electronics Engineers, Inc. Published 2020 by John Wiley & Sons, Inc.

supervisors exist. To improve the computational efficiency, Ghaffari *et al.* (2003) propose an intuitive optimal supervisor synthesis method. It first identifies a set of marking/transition separation instances (MTSIs), each of which is a pair of a legal marking and a transition whose firing at the marking leads the system to an illegal state. Then, Petri net monitors that contain control places and arcs are computed iteratively by using cycle equations, reachability conditions, and event separation conditions to prevent these transitions from firing at their corresponding markings and keep the system running in the legal zone. After that, Uzam and Zhou (2006) develop an iterative approach for Petri net supervisor synthesis for flexible manufacturing systems (FMSs), which is a typical kind of RASs. It divides the reachability graph of a plant net into a live zone (LZ) and deadlock zone (DZ). A marking in LZ is called a legal marking that can reach the initial marking, while a marking in DZ is a deadlock, a livelock, or a bad marking that inevitably leads to a deadlock or livelock. In DZ, first-met bad markings (FBMs) are those that are immediately reachable from some marking in LZ. At each iteration of the approach, an FBM is selected from the reachability graph. To prevent the FBM from being reached, a monitor is designed by constructing a place invariant according to the invariant-based control method in Yamalidou *et al.* (1996). Then, the monitor that contains a control place and several arcs is added to the plant net. This process is iteratively carried out until all FBMs are handled. Thus, the obtained final controlled net never enters DZ. However, the optimality of the added supervisor cannot be guaranteed since some markings in LZ may be forbidden by the controlled net. To address it, Chen *et al.* (2011) develop an approach that can obtain optimal Petri net supervisors for FMSs if such supervisors exist. It is also an iterative method that derives an optimal monitor by designing a place invariant on the control place and some activity places at each iteration, which is achieved by solving an integer linear program (ILP) to forbid as many FBMs as possible and permit all legal markings at the iteration. To alleviate the computational complexity problem, a vector covering method is developed to reduce the considered FBMs and legal markings. However, this method does not consider the minimization of a supervisor's structure, which is important to the control implementation of the underlying system. Hence, Chen and Li (2011) propose a supervisor synthesis method that formulates and solves a single integer linear program. The advantages of the method are that the obtained supervisor is not only behaviorally maximal but also structurally minimal in terms of the number of control places.

This chapter first gives the definitions of DP-nets that are used in the RAS supervisor synthesis approaches presented in this book. Then, it recalls the applications of the theory of regions to design optimal Petri net supervisors with the fewest monitors, including reachability graph analysis, supervisor computation with place invariants, and optimal and structure-minimal supervisor synthesis based on a vector covering method. Finally, an illustrative example and the application scope of the presented approach are given.

3.2 Petri Nets for Supervisory Synthesis

The methods in Chen *et al.* (2011), Chen and Li (2011), and Uzam and Zhou (2006) use the theory of regions that is a reachability graph-based analysis technique to synthesize Petri net supervisors for RASs. To handle deadlocks in RASs, several classes of Petri nets have been proposed in the literature, such as LS^3PR, S^3PR, ES^3PR, S^4PR, S^*PR, and S^5PR (the details of these nets can be seen in Section 2.2.5).The differences among these classes are mainly the number and types of resources that can be used by each operation and the structure of processing subnets. In addition, some of these classes are ordinary nets while others are generalized ones. For the ease of use in the supervisor synthesis, we unify them as the deadlock prevention nets (DP-nets) for RASs as follows.

Definition 3.1 *A DP-net $N = (P_0 \cup P_R \cup P_A, T, F, W)$, which models an RAS for deadlock prevention, is defined as the composition of a set of subnets $N^x = (\{p_0^x\} \cup P_R^x \cup P_A^x, T^x, F^x, W^x)$ that share some common places, i.e., $N = \bigcup_{x \in \mathbb{N}_J} N^x$ where $\mathbb{N}_J = \{i \in \mathbb{Z}^+ | N^i \text{ is the ith subnet of } N\}$ and the following statements are true.*

1) *p_0^x is called an idle place of N^x. Elements in P_A^x and P_R^x are called activity (operation) and resource places of N^x.*
2) *$P_A^x \neq \emptyset; P_R^x \neq \emptyset; p_0^x \notin P_A^x; (\{p_0^x\} \cup P_A^x) \cap P_R^x = \emptyset.$*
3) *$W = W_A \cup W_R$ with $W_A: F \cap ((P_A \cup P_0) \times T) \cup (T \times (P_A \cup P_0)) \to \{1\}$ and $W_R: F \cap ((P_R \times T) \cup (T \times P_R)) \to \mathbb{Z}^+.$*
4) *$\forall r \in P_R$, there exists a unique minimal P-semiflow I_r such that $\|I_r\| \cap P_R = \{r\}$, $\|I_r\| \cap P_0 = \emptyset$, $\|I_r\| \cap P_A \neq \emptyset$, and $I_r(r) = 1$. In addition, $P_A = \bigcup_{r \in P_R} (\|I_r\| \setminus \{r\}).$*
5) *$\forall r \in P_R^x, {}^\bullet r \cap r^\bullet = \emptyset.$*
6) *If the resource places in P_R^x and their connected arcs are removed from N^x, the resultant net $N^{x'}$ is a strongly connected state machine.*
7) *N^x and $N^y (x \neq y)$ are composable when they share a set of common places that are resource places.*
8) *Transitions in $p_0^{x\bullet}$ and ${}^\bullet p_0^x$ are respectively called source and sink transitions of N^x, which satisfy $p_0^{x\bullet\bullet} \cap P_R = {}^{\bullet\bullet}p_0^x \cap P_R = \emptyset.$*
9) *For any $r \in P_R$, the support of a minimal P-semiflow I_r without r, i.e., $H(r) = \|I_r\| \setminus \{r\}$, is called the set of holders of r, and $M_0(r) = \sum_{p \in \{r\} \cup H(r)} I_r(p) \cdot M(p) = C(r)$ is called the capacity of r.*

10) *For any $p \in P_A$, a $|P_R|$-dimensional vector $U_p(i), i = 1, \ldots, |P_R|$ is called the resource requirements of p, which indicates how many units of $r_i \in P_R$ are required by the operation of p.*

11) *An initial marking M_0 is called an acceptable one for N if i) $\forall p \in P_0$, $M_0(p) \geqslant 1$; ii) $\forall p \in P_A$, $M_0(p) = 0$; and iii) $\forall r \in P_R$, $M_0(r) \geqslant \max_{p \in \|r\|} I_r(p)$.*

In Definition 3.1, items 1 and 2 mean that the places of a DP-net are divided into three disjoint classes: idle, activity, and resource places. Item 3 indicates that the weight of any arc between an idle place (or an activity place) and a transition is not greater than one and the weight of an arc between a resource place and a transition can be greater than one. Item 4 delves into the resource preservation of DP-nets, explaining how only a resource place and some activity places hold and use the tokens of a special kind of resources. Item 11 gives the definitions of an acceptable initial marking for a DP-net. Most of PNs for RASs, such as LS^3PR, S^3PR, ES^3PR, S^4PR, S^*PR, and S^5PR, conform to the definitions of DP-nets, but PC^2R does not belong to DP-nets since the idle places in PC^2R also require resource tokens, which violates item 4.

Example 3.1   Consider an automated manufacturing system that is composed of two shared robots, R_1 and R_2, each of which holds one part at a time, one shared machine M_1 which processes one part at a time, two loading buffers I_1 and I_2, and two unloading buffers O_1 and O_2. The action areas of robot R_1 are I_1, M_1, R_2, and O_2, and the areas for robot R_2 are I_2, M_1, R_1, and O_1. As shown in Figure 3.1, two

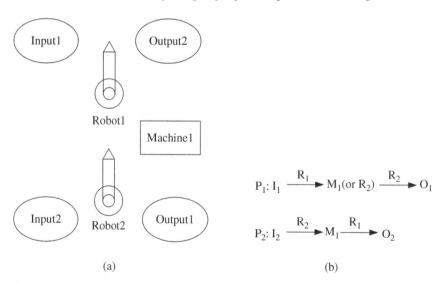

Figure 3.1 (a) A flexible manufacturing system; (b) its processing sequences.

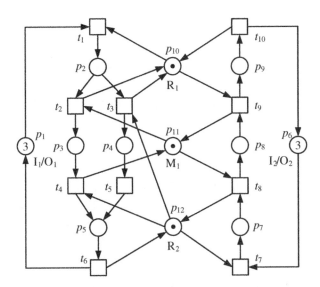

Figure 3.2 DP-net of the example RAS.

part types, P_1 and P_2, are processed in the system. P_1 is taken from I_1 by R_1, and after being handled by M_1, it is moved to O_1 by R_2. Note that in this system, P_1 can also be passed from R_1 to R_2 directly. P_2 is taken from I_2 by R_2, and after being handled by M_1, it is moved to O_2 by R_1. Assume that there are three parts of each part type to be processed.

Such a system can be modeled by a DP-net shown in Figure 3.2. It is a pure net with 10 transitions and 12 places in which $P_0 = \{p_1, p_6\}$, $P_A = \{p_2 - p_5, p_7 - p_9\}$, and $P_R = \{p_{10} - p_{12}\}$. For the resource place p_{10}, its minimal P-semiflow is $I_{p_{10}} = (0\ 1\ 0\ 0\ 0\ 0\ 0\ 0\ 1\ 1\ 0\ 0)^T = p_2 + p_9 + p_{10}$ such that $\|I_{p_{10}}\| \cap P_R = \{p_{10}\}$, $\|I_{p_{10}}\| \cap P_0 = \emptyset$, $\|I_{p_{10}}\| \cap P_A \neq \emptyset$, and $I_{p_{10}}(p_{10}) = 1$. In addition, if all resource places and their connected arcs are removed, the resultant subnets are two strongly connected state machines.

Note that DP-nets are generalized nets since the weight of an arc between a resource place and a transition can be greater than one.

3.3 Optimal and Minimal Supervisory Synthesis

3.3.1 Reachability Graph Analysis

Given a DP-net, its reachability graph $G(N, M_0)$ consists of two disjointed parts: LZ and DZ. The markings in LZ are legal ones that can reach the initial marking M_0. The set of legal markings is defined as:

$$\mathcal{M}_L = \{M | M \in R(N, M_0) \wedge M_0 \in R(N, M)\}. \tag{3.1}$$

DZ consists of all deadlock markings, livelock markings, and bad markings that inevitably lead to deadlocks and livelocks, which are called illegal markings. An FBM is a marking in DZ, which is the very first marking from LZ to DZ. The set of FBMs is defined as:

$$\mathcal{M}_F = \{M \in DZ | \exists M' \in LZ, t \in T, \text{ s.t. } M'[t\rangle M\}. \tag{3.2}$$

For the DP-net in Figure 3.2, its reachability graph is given in Figure 3.3 where a compact multiset formalism $\sum_i M(p_i)p_i$ is used to denote each marking of the net for conciseness. In the reachability graph, $M_9 = p_1 + p_2 + p_3 + 2p_6 + p_7, M_{11} = 2p_1 + p_3 + 2p_6 + p_7 + p_{10}$, and $M_{22} = 2p_1 + p_2 + p_6 + p_7 + p_8$ are FBMs while the others are legal markings. Note that M_9 and M_{22} are also deadlock markings. If all FBMs are forbidden and all legal markings are permitted by a designed supervisor, the resultant controlled net keeps running in LZ but never enters DZ.

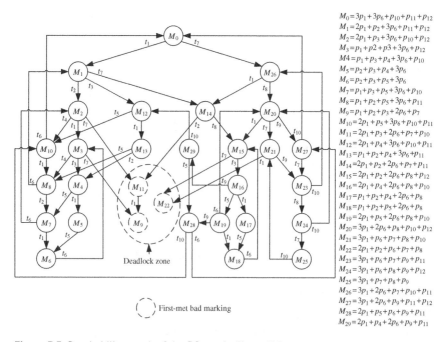

$M_0 = 3p_1 + 3p_6 + p_{10} + p_{11} + p_{12}$
$M_1 = 2p_1 + p_2 + 3p_6 + p_{11} + p_{12}$
$M_2 = 2p_1 + p_3 + 3p_6 + p_{10} + p_{12}$
$M_3 = p_1 + p2 + p3 + 3p_6 + p_{12}$
$M4 = p_1 + p_3 + p_4 + 3p_6 + p_{10}$
$M_5 = p_2 + p_3 + p_4 + 3p_6$
$M_6 = p_2 + p_3 + p_5 + 3p_6$
$M_7 = p_1 + p_3 + p_5 + 3p_6 + p_{10}$
$M_8 = p_1 + p_2 + p_5 + 3p_6 + p_{11}$
$M_9 = p_1 + p_2 + p_3 + 2p_6 + p_7$
$M_{10} = 2p_1 + p_5 + 3p_6 + p_{10} + p_{11}$
$M_{11} = 2p_1 + p_3 + 2p_6 + p_7 + p_{10}$
$M_{12} = 2p_1 + p_4 + 3p_6 + p_{10} + p_{11}$
$M_{13} = p_1 + p_2 + p_4 + 3p_6 + p_{11}$
$M_{14} = 2p_1 + p_2 + 2p_6 + p_7 + p_{11}$
$M_{15} = 2p_1 + p_2 + 2p_6 + p_8 + p_{12}$
$M_{16} = 2p_1 + p_4 + 2p_6 + p_8 + p_{10}$
$M_{17} = p_1 + p_2 + p_4 + 2p_6 + p_8$
$M_{18} = p_1 + p_2 + p_5 + 2p_6 + p_8$
$M_{19} = 2p_1 + p_5 + 2p_6 + p_8 + p_{10}$
$M_{20} = 3p_1 + 2p_6 + p_8 + p_{10} + p_{12}$
$M_{21} = 3p_1 + p_6 + p_7 + p_8 + p_{10}$
$M_{22} = 2p_1 + p_2 + p_6 + p_7 + p_8$
$M_{23} = 3p_1 + p_6 + p_7 + p_9 + p_{11}$
$M_{24} = 3p_1 + p_6 + p_8 + p_9 + p_{12}$
$M_{25} = 3p_1 + p_7 + p_8 + p_9$
$M_{26} = 3p_1 + 2p_6 + p_7 + p_{10} + p_{11}$
$M_{27} = 3p_1 + 2p_6 + p_9 + p_{11} + p_{12}$
$M_{28} = 2p_1 + p_5 + p_6 + p_9 + p_{11}$
$M_{29} = 2p_1 + p_4 + 2p_6 + p_9 + p_{11}$

Figure 3.3 Reachability graph of the DP-net in Figure 3.2.

3.3.2 Supervisor Computation with Place Invariants

This subsection presents a method to compute a Petri net supervisor that comprises control places and arcs via a place invariant-based approach. Let $[N]$ be the incidence matrix of a controlled net after adding a supervisor. It contains the information from a plant net and the supervisor, i.e.,

$$[N] = \begin{bmatrix} N_p \\ N_c \end{bmatrix} \tag{3.3}$$

where $[N_p]$ denotes the incidence matrix of the plant net and $[N_c]$ is the incidence matrix between control places and the transitions in the plant net.

The supervisor is designed via some place invariants as follows:

$$[L] \cdot M_p + M_c = b \tag{3.4}$$

where $[L]$ is an $n_c \times n$ non-negative integer matrix that represents a coefficient matrix of the designed place invariants, M_p is an $n \times 1$ marking vector of the plant net, M_c is an $n_c \times 1$ marking vector of the control places, b is an $n_c \times 1$ non-negative integer vector, n_c is the number of control places in the supervisor, and n is the number of places in the plant net. Since M_c is not negative, we have:

$$[L] \cdot M_p \leqslant b, \tag{3.5}$$

which should be satisfied by the controlled net.

The supervisor $[N_c]$ can be computed as:

$$[N_c] = -[L] \cdot [N_p] \tag{3.6}$$

where $[N_c](i,j) > 0$ indicates that there is an added arc from transition $t_j \in T$ to control place p_{c_i} such that $W(t_j, p_{c_i}) = [N_c](i,j)$ and $[N_c](i,j) < 0$ implies an arc from p_{c_i} to $t_j \in T$ such that $W(p_{c_i}, t_j) = -[N_c](i,j)$.

Let M_{c_0} be the initial marking of the control places. It should also satisfy (3.4). Thus, M_{c_0} can be computed by the following equation:

$$M_{c_0} = b - [L] \cdot M_{p_0}. \tag{3.7}$$

3.3.3 Optimal Supervisor Synthesis and Vector Covering Method

Places in a DP-net model are classified into three types: idle, resource, and activity places. Initial tokens in an idle place represent the maximal concurrent operations that can be performed in a processing subnet. Resource places denote resources required by operations, e.g., robots and machines, and their initial tokens represent available resource units at the initial marking. Activity places represent operations to be performed to parts with some resources, and they have no token at the initial marking. For such a net, only activity places need to be considered to design place invariants in supervisor synthesis (Uzam & Zhou, 2007).

Definition 3.2 *A supervisor is said to be optimal (maximally permissive) if all legal markings are permitted and all illegal markings are forbidden by the controlled net with the supervisor.*

An optimal supervisor can be obtained by designing some place invariants that permit all legal markings and forbid all FBMs since any path from LZ (legal markings) to DZ (illegal markings) goes through an FBM. An FBM $M_j \in \mathcal{M}_F$ is forbidden by a designed place invariant if

$$\sum_{i \in \mathbb{N}_A} l_i \cdot M(p_i) \leqslant \beta \tag{3.8}$$

where $\mathbb{N}_A = \{i | p_i \in P_A\}$,

$$\beta = \sum_{i \in \mathbb{N}_A} l_i \cdot M_j(p_i) - 1, \tag{3.9}$$

and $l_i (i \in \mathbb{N}_A)$ is the ith coefficient of the place invariant.

All legal markings in LZ are permitted by the place invariant if:

$$\sum_{i \in \mathbb{N}_A} l_i \cdot M_m(p_i) \leqslant \beta, \ \forall M_m \in \mathcal{M}_L. \tag{3.10}$$

For an FBM M_j, by substituting (3.9) into (3.10), the constraint to forbid M_j and survive all legal markings becomes

$$\sum_{i \in \mathbb{N}_A} l_i \cdot \left(M_m(p_i) - M_j(p_i) \right) \leqslant -1, \ \forall M_m \in \mathcal{M}_L. \tag{3.11}$$

Constraint (3.11) is called a reachability constraint and determines the coefficients $l_i (i \in \mathbb{N}_A)$ of the place invariant. Thus, for FBM M_j, if $l_i (i \in \mathbb{N}_A)$ of a place invariant satisfies (3.8) and (3.11), then the place invariant forbids M_j and permits all legal markings. In this case, the corresponding monitor, which contains a control place and several arcs, is optimal since all legal markings of the plant net are reachable after adding the monitor.

However, the numbers of legal markings and FBMs in $G(N, M_0)$ are often large since the size of $G(N, M_0)$ grows exponentially with the Petri net size. To handle it, a vector covering method is used to reduce the number of markings to be considered in the supervisor synthesis.

Definition 3.3 $\forall M, M' \in R(N, M_0), M \geqslant_A M'$ if $\forall p \in P_A, M(p) \geqslant M'(p)$.

According to (3.8), if $M \geqslant_A M'$ and M' is forbidden by a monitor, then M is also forbidden by the monitor. Similarly, according to (3.11), if $M \geqslant_A M'$ and M is permitted by a monitor, then M' is also permitted by the monitor.

Definition 3.4 $\mathcal{M}_F^*$ *is called a minimal covered set of* $\mathcal{M}_F$ *if*

1) $\mathcal{M}_F^* \subseteq \mathcal{M}_F$;
2) $\forall M \in \mathcal{M}_F, \exists M' \in \mathcal{M}_F^*$ *such that* $M \geqslant_A M'$; *and*
3) $\forall M \in \mathcal{M}_F^*, \nexists M'' \in \mathcal{M}_F^*$ *such that* $M \geqslant_A M''$ *and* $M \neq M''$.

Definition 3.5 $\mathcal{M}_L^*$ *is called a minimal covering set of* $\mathcal{M}_L$ *if*

1) $\mathcal{M}_L^* \subseteq \mathcal{M}_L$;
2) $\forall M \in \mathcal{M}_L, \exists M' \in \mathcal{M}_L^*$ *subject to* $M' \geqslant_A M$; *and*
3) $\forall M \in \mathcal{M}_L^*, \nexists M'' \in \mathcal{M}_L^*$ *subject to* $M'' \geqslant_A M$ *and* $M \neq M''$.

If a monitor or a supervisor forbids all markings in $\mathcal{M}_F^*$, it also forbids all FBMs in $\mathcal{M}_F$. If a monitor or a supervisor permits all markings in $\mathcal{M}_L^*$, it also permits all legal markings in $\mathcal{M}_L$. Thus, only two reduced sets, $\mathcal{M}_F^*$ and $\mathcal{M}_L^*$, should be considered when designing an optimal supervisor to forbid all FBMs and permit all legal markings of the original net. Therefore, given FBM M_j, the reachability constraint that forbids M_j and survives all legal markings via a place invariant is reduced to:

$$\sum_{i \in \mathbb{N}_A} l_i \cdot \left(M_m(p_i) - M_j(p_i) \right) \leqslant -1, \ \forall M_m \in \mathcal{M}_L^*. \tag{3.12}$$

3.3.4 Optimal Supervisor with Fewest Monitors

A designed place invariant may forbid more than one FBM. Given a place invariant I_j designed to forbid $M_j \in \mathcal{M}_F^*$, there exists $M_k \in \mathcal{M}_F^*(k \neq j)$ which is also forbidden by I_j if:

$$\sum_{i \in \mathbb{N}_A} l_{j,i} \cdot \left(M_k(p_i) - M_j(p_i) \right) \geqslant -\Gamma \cdot (1 - f_{j,k}),$$

$$\forall M_k \in \mathcal{M}_F^* \text{ and } k \neq j \tag{3.13}$$

where $l_{j,i}$ is a coefficient of place invariant I_j, Γ is a positive integer that is chosen big enough, and $f_{j,k} \in \{0,1\}$ where $f_{j,k} = 1$ indicates that place invariant I_j designed to forbid M_j also forbids M_k and $f_{j,k} = 0$ indicates that I_j does not forbid M_k.

A set of variables q_j ($j \in \mathbb{N}_F^* = \{i | M_i \in \mathcal{M}_F^*\}$) for I_j is introduced such that:

$$f_{j,k} \leqslant q_j, \ \forall j, k \in \mathbb{N}_F^* \text{ and } k \neq j \tag{3.14}$$

where $q_j \in \{0, 1\}$. If $q_j = 1$, it means that I_j is selected to design a monitor in the supervisor; if $q_j = 0$, it means that I_j is not. Constraints (3.14) indicate that, only when I_j is selected, can it forbid M_j as well as the FBMs satisfying (3.13). To ensure that all FBMs in $\mathcal{M}_F^*$ are forbidden by the designed place invariants, $f_{k,j}$ and q_j should satisfy

$$q_j + \sum_{k \in \mathbb{N}_F^*, k \neq j} f_{k,j} \geqslant 1. \tag{3.15}$$

Combining the above constraints and an objective to minimize the number of designed monitors, we obtain the following ILP to design an optimal supervisor with the fewest monitors, which is denoted as the minimal number of monitor problem (MMP).

$$\min \sum_{j \in \mathbb{N}_F^*} q_j \tag{3.16}$$

subject to

$$\sum_{i \in \mathbb{N}_A} l_{j,i} \cdot \left(M_m(p_i) - M_j(p_i) \right) \leqslant -1,$$

$$\forall M_j \in \mathcal{M}_F^* \text{ and } \forall M_m \in \mathcal{M}_L^* \tag{3.17}$$

$$\sum_{i \in \mathbb{N}_A} l_{j,i} \cdot \left(M_k(p_i) - M_j(p_i) \right) \geqslant -\Gamma \cdot (1 - f_{j,k}),$$

$$\forall M_j, M_k \in \mathcal{M}_F^* \text{ and } j \neq k \tag{3.18}$$

$$f_{j,k} \leqslant q_j, \forall j, k \in \mathbb{N}_F^* \text{ and } j \neq k \tag{3.19}$$

$$q_j + \sum_{k \in \mathbb{N}_F^*, k \neq j} f_{k,j} \geqslant 1, \forall j \in \mathbb{N}_F^* \tag{3.20}$$

$$l_{j,i} \in \{0, 1, 2, \dots\}, \forall i \in \mathbb{N}_A \text{ and } \forall j \in \mathbb{N}_F^*$$

$$f_{j,k} \in \{0, 1\}, \forall j, k \in \mathbb{N}_F^* \text{ and } j \neq k$$

$$q_j \in \{0, 1\}, \forall j \in \mathbb{N}_F^*$$

Note that the coefficients $l_{j,i}$ ($j \in \mathbb{N}_F^*$ and $i \in \mathbb{N}_A$) of place invariants are nonnegative and the MMP minimizes its objective under this restriction.

3.3.5 Deadlock Prevention Policy

This section presents a deadlock prevention policy by using MMP to obtain an optimal supervisor with the fewest monitors.

Algorithm 3.1

```
Input: A DP-net system (N, M₀) of an RAS.
Output: A maximally permissive net with the fewest monitors.
Deadlock_prevention {
        Generate G(N, M₀) and compute 𝓜_F and 𝓜_L;
        Compute 𝓜*_F and 𝓜*_L via the vector covering method in Section
3.3.3;
        V_P := ∅, V_A := ∅; /* V_P and V_A denote the sets of control
places and added arcs in a supervisor. */
        Solve the MMP given in Section 3.3.4. If it has no solution,
exit;
        For each q_j = 1
        {
            Use l_{j,i} in the solution as the coefficient of a place
invariant and design a control place p_{c_j} and the arcs A_j associ-
ated with p_{c_j} by the method presented in Section 3.3.2;
            V_P := V_P ∪ {p_{c_j}}, V_A := V_A ∪ {A_j};
        }
        Add V_P and V_A to (N, M₀) and output the resulting net.
}
```

First, Algorithm 3.1 generates all markings in $\mathcal{M}_F$ and $\mathcal{M}_L$ of $G(N, M_0)$. Next, it computes $\mathcal{M}_F^*$ and $\mathcal{M}_L^*$ by using the vector covering method. Then, MMP is formulated and solved to compute the control places and arcs in the supervisor. Finally, the controlled net is obtained by adding the supervisor to the original net. In the following, we prove that Algorithm 3.1 obtains an optimal supervisor with the fewest monitors for the DP-nets of RASs if such a supervisor exists.

Theorem 3.1 *Assume that each monitor obtained by Algorithm 3.1 is associated with a P-semiflow. Algorithm 3.1 obtains an optimal supervisor with the fewest monitors for the DP-net of an RAS if and only if MMP has an optimal solution.*

Proof: *First, we prove that if MMP has an optimal solution, Algorithm 3.1 obtains an optimal supervisor with the fewest monitors with the condition that each monitor is associated with a P-semiflow. Theorem 6 in Chen et al. (2011) has shown that each monitor is associated with a place invariant. Since the coefficient $l_{j,i}$ ($j \in \mathbb{N}_F^*$ and $i \in \mathbb{N}_A$) of any place invariant in MMP is non-negative, the place invariants designed by Algorithm 3.1 that solves MMP are P-semiflows. According to (3.20), MMP forbids all FBMs since $\forall M \in \mathcal{M}_F^*$, M is forbidden by at least one place invariant. On the other hand, MMP permits all legal markings in $G(N, M_0)$ by (3.17). Thus, the obtained supervisor is optimal, i.e., behaviorally maximal. The objective of MMP ensures that the number of added monitors is minimized. Therefore, the obtained supervisor is optimal and has the fewest monitors.*

Then, we aim to prove that if an optimal supervisor exists with the premise that each monitor of the supervisor is associated with a P-semiflow, then MMP has an optimal solution. In the following, we first prove that any designed P-semiflow for an optimal control purpose satisfies (3.17). Suppose that there exists a P-semiflow that does not satisfy (3.17) for some legal markings in $\mathcal{M}_L^*$. Then, as Section 3.3.3 describes, the P-semiflow forbids these legal markings. So, the control place that corresponds to this P-semiflow is not optimal. By contradiction, it is proven that any P-semiflow designed to obtain an optimal supervisor satisfies (3.17). Thus, for each FBM, there exists a P-semiflow satisfying (3.17) such that it forbids the FBM and permits all legal markings. Therefore, MMP has a solution. In addition, since MMP has an objective function to minimize the number of added monitors, MMP has an optimal solution. ∎

3.4 An Illustrative Example

In this section, we use the DP-net shown in Figure 3.2 to illustrate the presented approach. Its reachability graph is given in Figure 3.3. It has 30 reachable markings in which 27 are legal and 3 are FBMs. By using the vector covering method, we have $M_F^* = \{p_3 + p_7, p_2 + p_7 + p_8\}$ and $M_L^* = \{p_2 + p_3 + p_4, p_2 + p_3 + p_5, p_2 + p_7, p_2 + p_4 + p_8, p_2 + p_5 + p_8, p_7 + p_8 + p_9, p_5 + p_9, p_4 + p_9\}$.

According to (3.17), $\forall M_j \in M_F^*$, there exists a place invariant I_j that forbids M_j and permits all legal markings in the original net if the following constraints are satisfied.

$$l_{1,1} + l_{1,3} - l_{1,5} \leqslant -1 \tag{3.21}$$

$$l_{1,1} + l_{1,4} - l_{1,5} \leqslant -1 \tag{3.22}$$

$$l_{1,1} - l_{1,2} \leqslant -1 \tag{3.23}$$

$$l_{1,1} - l_{1,2} + l_{1,3} - l_{1,5} + l_{1,6} \leqslant -1 \tag{3.24}$$

$$l_{1,1} - l_{1,2} + l_{1,4} - l_{1,5} + l_{1,6} \leqslant -1 \tag{3.25}$$

$$-l_{1,2} + l_{1,6} + l_{1,7} \leqslant -1 \tag{3.26}$$

$$-l_{1,2} + l_{1,4} - l_{1,5} + l_{1,7} \leqslant -1 \tag{3.27}$$

$$-l_{1,2} + l_{1,3} - l_{1,5} + l_{1,7} \leqslant -1 \tag{3.28}$$

$$l_{2,2} + l_{2,3} - l_{2,5} - l_{2,6} \leqslant -1 \tag{3.29}$$

$$l_{2,2} + l_{2,4} - l_{2,5} - l_{2,6} \leqslant -1 \tag{3.30}$$

$$-l_{2,6} \leqslant -1 \tag{3.31}$$

$$l_{2,3} - l_{2,5} \leqslant -1 \tag{3.32}$$

$$l_{2,4} - l_{2,5} \leqslant -1 \tag{3.33}$$

$$-l_{2,1} + l_{2,7} \leqslant -1 \tag{3.34}$$

$$-l_{2,1} + l_{2,4} - l_{2,5} - l_{2,6} + l_{2,7} \leqslant -1 \tag{3.35}$$

$$-l_{2,1} + l_{2,3} - l_{2,5} - l_{2,6} + l_{2,7} \leqslant -1 \tag{3.36}$$

Note that only activity places are considered in supervisor synthesis. Next, for PI I_j designed to forbid $M_j \in \mathcal{M}_F^*$, there may exist $M_k \in \mathcal{M}_F^*(k \neq j)$ that is also forbidden by I_j. So, we have:

$$l_{1,1} - l_{1,2} + l_{1,6} \geqslant -\Gamma \cdot (1 - f_{1,2}) \tag{3.37}$$

$$-l_{2,1} + l_{2,2} - l_{2,6} \geqslant -\Gamma \cdot (1 - f_{2,1}) \tag{3.38}$$

Then, a PI designed to forbid an FBM in $\mathcal{M}_F^*$ can forbid another FBM in M_F^* only when it is selected to design a monitor. It can be described below.

$$f_{1,2} \leqslant q_1 \tag{3.39}$$

$$f_{2,1} \leqslant q_2 \tag{3.40}$$

After that, the following constraints should be satisfied to ensure that each FBM in $\mathcal{M}_F^*$ is forbidden by at least one place invariant.

$$q_1 + f_{2,1} \geqslant 1 \tag{3.41}$$

$$q_2 + f_{1,2} \geqslant 1 \tag{3.42}$$

Finally, we minimize the number of monitors in the designed supervisor, which means that the number of selected place invariants should be minimal. So, the objective function of the ILP is:

$$\min \sum_{j \in \mathbb{N}_F^*} q_j \tag{3.43}$$

By solving the inequalities (3.22)–(3.42) with the objective (3.43), we obtain an optimal solution with $q_1 = f_{1,2} = 1$, $l_{1,1} = l_{1,6} = l_{2,1} = l_{2,2} = l_{2,5} = l_{2,6} = 1$, $l_{1,2} = l_{1,5} = 2$, and all other variables equalling zero. Since $q_1 = 1$, a control place p_{c_1} is designed for place invariant I_1: $p_2 + 2p_3 + 2p_7 + p_8 + p_{c_1} = b$ as described in (3.4). According to the supervisor computation method in Section 3.3.2, we can obtain the preset, postset, and initial marking of p_{c_1}, i.e., $\cdot p_{c_1} = \{t_3, 2t_4, t_8, t_9\}$, $p_{c_1}^\bullet = \{t_1, t_2, 2t_7\}$, and $M_0(p_{c_1}) = b = 3$. Thus, a control place with three initial

Table 3.1 Supervisor obtained for the DP-net in Figure 3.2.

i	FBM$_i$	l_i	$^\bullet p_{c_i}$	$p_{c_i}^\bullet$	$M_0(p_{c_i})$
1	$p_3 + p_7$	$p_2 + 2p_3 + 2p_7 + p_8 \leqslant 3$	$t_3, 2t_4, t_8, t_9$	$t_1, t_2, 2t_7$	3

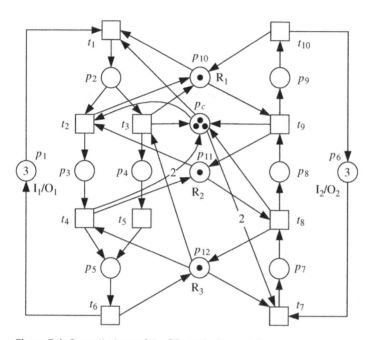

Figure 3.4 Controlled net of the DP-net in Figure 3.2.

tokens and seven connected arcs are obtained in the supervisor. Table 3.1 shows the details of the results. Adding the supervisor to the original net, we obtain a controlled net shown in Figure 3.4 that permits all 27 legal markings of the original net and has no deadlocks, livelocks, or bad markings that inevitably lead to deadlocks and livelocks. In addition, it guarantees that the obtained controlled net is behaviorally maximal and the added supervisor has the fewest monitors according to Theorem 3.1.

3.5 Concluding Remarks

For deadlock prevention of RASs, this chapter presents an application of the theory of regions in designing an optimal Petri net supervisor with the fewest monitors by formulating and solving a single ILP. The performance of a deadlock prevention policy can be evaluated by three criteria: behavioral permissiveness, structural complexity, and computational complexity. The general advantages of the presented approach are that its controlled nets are maximally permissive in behavior and the added supervisors are structurally minimal in terms of monitors. It can be applied to most of the Petri nets for RASs in the literature, such as S^3PR, ES^3PR, LS^3PR, ELS^3PR, GLS^3PR, WS^3PR, S^3PGR^2, WS^3PSR, S^4R, S^4PR, S^*PR, and S^5PR whose definitions can be seen in Section 2.2.5.

However, the approach has high computational complexity since it needs to generate the whole reachability graph of a Petri net and solve an ILP. The number of reachable markings in the graph increases exponentially with the net size and the number of constraints in the ILP is polynomial with respect to the size of the minimal covering set of legal markings and the minimal covered set of FBMs. Therefore, this approach can only be applied to small-scale systems. In the following chapters, some acceleration strategies that speed up the supervisor synthesis process are introduced.

3.6 Bibliographical Notes

The classification of places in Petri net models of RASs can be found in Ezpeleta *et al.* (1995), Park and Reveliotis (2001), Tricas *et al.* (1995), and Zhou and DiCesare (1991). The well-known supervisor computation method based on place invariants is presented in Yamalidou *et al.* (1996). Uzam (2002) is the first to use the theory of regions to synthesize Petri net supervisors and Uzam & Zhou (2006) give the concept of FBMs. Then, Ghaffari *et al.* (2003) propose a popular linear algebraic approach to design optimal Petri net supervisors. For more details about the ILP formulation and the vector covering method, please refer to Chen and Li (2011). Li *et al.* (2008*a*) give a novel method to combine siphon analysis with the theory of regions to reduce the computational burdens of supervisor synthesis. An improved version of the theory of regions can be found in Huang *et al.* (2012*b*). Further work has been done in Huang et al. (2019).

4

Supervisor Design with Fewer Places

Solving a state separation problem (SSP) is an important technique used to obtain liveness-enforcing or deadlock-free supervisors for resource allocation systems (RASs) based on reachability graph analysis. It partitions a reachability graph into a live zone and a deadlock zone. First-met bad markings, which exist in the deadlock zone and are the very first entries from the live zone to the deadlock zone, are forbidden by designed place invariants to form a supervisor, preventing the system from entering the deadlock zone. In addition, the designed place invariants permit as many markings in the live zone as possible to allow high permissiveness of the supervisor. This chapter presents an approach to reduce the number of places considered in such place invariant designs. First, the concepts of critical transitions and critical activity places are introduced. Also, an algorithm used to quickly compute them is given. Then, the proof that only critical activity places need to be considered in such place invariant designs is established. The approach can reduce the number of places considered in the supervisor synthesis. As a result, SSP becomes simpler.

4.1 Introduction

Methods to synthesize Petri net supervisors based on reachability graph analysis can mainly be divided into two categories: the event/state separation problem (ESSP) and the state separation problem (SSP).

The ESSP methods (Chen *et al.*, 2014*b*; Ghaffari *et al.*, 2003; Huang *et al.*, 2012*b*; Li *et al.*, 2008*a*; Uzam, 2002) identify the set of state/event separation instances, also called marking/transition separation instances (MTSIs). Each state/event separation instance is a pair that includes a marking in the live zone and a transition in a way that the firing of the transition at the marking leads to a marking in

Supervisory Control and Scheduling of Resource Allocation Systems: Reachability Graph Perspective,
First Edition. Bo Huang and MengChu Zhou.

the deadlock zone. Some monitors, each of which consists of a control place and some connected arcs, are computed based on cycle equations, reachability conditions, and event separation conditions, preventing the transitions in MTSIs from firing at their corresponding markings and keeping the system running in the live zone.

The SSP methods (Chen & Li, 2011; Cordone & Piroddi, 2011; Huang *et al.*, 2015*b*; Uzam & Zhou, 2007) are based on some designed place invariants, each of which consists of a set of places whose weighted sum of tokens remain constant at any reachable marking. An objective of the SSP methods is to prevent all first-met bad markings (FBMs), which are some markings in the deadlock zone and immediately reachable from markings in the live zone, from being reached in the controlled net by constructing some place invariants. Yamalidou *et al.* (1996) first propose a well-established method for supervisor synthesis based on place invariants. Based on it, Uzam and Zhou (2007) develop an iterative method to construct a place invariant that forbids a selected FBM at each iteration. When the iterative method terminates, all FBMs are forbidden by the constructed place invariants and a liveness-enforcing and near-optimal supervisor is obtained for the underlying system.

To ensure the supervisor's optimality, the methods in Chen and Li (2011), Cordone and Piroddi (2011), and Huang *et al.* (2015*b*) construct the place invariants via a single integer linear program (ILP) to forbid all FBMs and permit all legal markings. However, the size of such ILP is an inherent barrier to apply the methods to large-scale Petri nets since the number of inequalities in it increases exponentially with the net size. To simplify ILP, Uzam and Zhou (2006) find that only activity (operation) places are needed to construct place invariants to prevent FBMs from being reached and the fewer places considered in place invariant construction, the simpler the method. However, the number of inequalities in ILPs is still large for median-size problems. Thus, further reducing the number of places to be considered in the place invariant design is important to accelerate a supervisor synthesis process.

This chapter presents a method that addresses this problem. First, the concepts of critical transitions and critical activity places for the DP-nets of RASs are introduced. Then, an algorithm used to quickly compute critical and free activity places is given. Next, it comments on two place invariant design approaches (Chen & Li, 2011; Uzam & Zhou, 2007) and proves that only critical activity places need to be considered in the construction of place invariants that forbid all FBMs and/or permit all legal markings. Finally, an algorithm to obtain supervisors via the place invariants on critical activity places is given.

4.2 Critical and Free Activity Places

This section introduces the definitions of critical and free activity places, to be used by the method presented in this chapter, and an algorithm to quickly identify them in a DP-net for RASs.

According to Definition 3.1, in a DP-net $N = (P_0 \cup P_R \cup P_A, T, F, W) = \bigcup_{x \in \mathbb{N}_J} N^x$, $\forall t \in T$, we have $\exists x' \in \mathbb{N}_J, t \in N^{x'}$ and $N^{x'}$ is a strongly connected state machine, i.e., $|{}^{\bullet}t| = |t^{\bullet}| = 1$. Then, the following property of DP-nets is implied.

Property 4.1 *Let $[N_{P_A}]$ be the incidence matrix of P_A and T of a DP-net $N = (P_0 \cup P_R \cup P_A, T, F, W)$. There are three cases for a column in $[N_{P_A}]$: i) one entry is 1 and the rest are 0, ii) one entry is -1 and the rest are 0, and iii) two entries are 1 and -1, respectively, and the rest are 0.*

In a net system (N, M_0) of a DP-net N, a transition $t \in T$ is enabled at a marking $M \in R(N, M_0)$ if $\forall p \in {}^{\bullet}t : M(p) \geqslant W(p, t)$. This is denoted by $M[\overrightarrow{t}\rangle$. If $M[\overrightarrow{t}\rangle$, then firing t at M yields a marking M' such that

$$M'(p) = \begin{cases} M(p) - W(p, t), & p \in {}^{\bullet}t \\ M(p) + W(t, p), & p \in t^{\bullet} \\ M(p), & \text{otherwise.} \end{cases} \tag{4.1}$$

It is denoted as $M[\overrightarrow{t}\rangle M'$ and M' is said to be reachable from M. If there exist a sequence of transitions $\sigma = t_1 t_2 \dots t_n$ and markings $M_1, M_2, \dots,$ and M_{n-1} such that $M[\overrightarrow{t_1}\rangle M_1[\overrightarrow{t_2}\rangle M_2 \dots M_{n-1}[\overrightarrow{t_n}\rangle M'$, we denote it as $M[\overrightarrow{\sigma}\rangle M'$. Next we define conversely enabling and firing a transition.

Definition 4.1 *A transition t is called conversely enabled at a marking M if $\forall p \in t^{\bullet} : M(p) \geqslant W(t, p)$, which is denoted as $M[\overleftarrow{t}\rangle$.*

Definition 4.2 *Suppose that a marking $M \in R(N, M_0)$ and a transition $t \in T$ satisfy $M[\overleftarrow{t}\rangle$. Conversely firing t at M yields a marking M' such that*

$$M'(p) = \begin{cases} M(p) + W(p, t), & p \in {}^{\bullet}t \\ M(p) - W(t, p), & p \in t^{\bullet} \\ M(p), & \text{otherwise.} \end{cases} \tag{4.2}$$

It is denoted as $M[\overleftarrow{t}\rangle M'$ and M' is said to be conversely reachable from M. In addition, if there exist a sequence of transitions $\sigma = t_1 t_2 \dots t_n$ and markings $M_1, M_2, \dots,$ and M_{n-1} such that $M[\overleftarrow{t_1}\rangle M_1[\overleftarrow{t_2}\rangle M_2 \dots M_{n-1}[\overleftarrow{t_n}\rangle M'$, we denote it as $M[\overleftarrow{\sigma}\rangle M'$.

It is known that places in a DP-net are classified into idle places P_0, resource places P_R, and activity places P_A. In the following, the places and transitions of a DP-net are further classified into more subclasses.

Definition 4.3 *Let $N = (P_0 \cup P_R \cup P_A, T, F, W)$ be a DP-net. P_R is partitioned into two disjoint subsets: R_S and $R_{\bar{S}}$. $R_S = \{p \in P_R | \, |H(r)| \geqslant 2\}$ is called the set of shared resource places and $R_{\bar{S}} = P_R \backslash R_S$ is called the set of unshared resource places.*

Definition 4.4 *Let $N = (P_0 \cup P_R \cup P_A, T, F, W) = \bigcup_{x \in \mathbb{N}_J} N^x$ be a DP-net. $\forall x \in \mathbb{N}_J$, P_A^x is partitioned into two disjoint subsets: A_K^x, the set of critical activity places in N^x, and A_F^x, the set of free activity places in N^x. A_F^x is further divided into $A_{\vec{F}}^x$, the set of forward free activity places in N^x, and $A_{\overleftarrow{F}}^x$, the set of backward free activity places in N^x. Accordingly, P_A is partitioned into two disjoint parts: the set of critical activity places A_K and the set of free activity places A_F such that $A_K = \bigcup_{x \in \mathbb{N}_J} A_K^x$ and $A_F = \bigcup_{x \in \mathbb{N}_J} A_F^x$. A_F is further divided into two subsets: $A_{\vec{F}} = \bigcup_{x \in \mathbb{N}_J} A_{\vec{F}}^x$ and $A_{\overleftarrow{F}} = \bigcup_{x \in \mathbb{N}_J} A_{\overleftarrow{F}}^x$.*

Definition 4.5 *Let $N = (P_0 \cup P_R \cup P_A, T, F, W) = \bigcup_{x \in \mathbb{N}_J} N^x$ be a DP-net. $\forall x \in \mathbb{N}_J$, T^x is partitioned into two disjoint subsets: T_K^x, the set of critical transitions in N^x, and T_F^x, the set of free transitions in N^x. T_F^x is further divided into $T_{\vec{F}}^x$, the set of forward free transitions in N^x, and $T_{\overleftarrow{F}}^x$, the set of backward free transitions in N^x. Accordingly, T is partitioned into two disjoint parts: the set of critical transitions T_K and the set of free transitions T_F such that $T_K = \bigcup_{x \in \mathbb{N}_J} T_K^x$ and $T_F = \bigcup_{x \in \mathbb{N}_J} T_F^x$. T_F is further divided into two subsets: $T_{\vec{F}} = \bigcup_{x \in \mathbb{N}_J} T_{\vec{F}}^x$ and $T_{\overleftarrow{F}} = \bigcup_{x \in \mathbb{N}_J} T_{\overleftarrow{F}}^x$.*

Definition 4.6 *$T_{\vec{F}}$ and $A_{\vec{F}}$ are recursively defined as: $\forall p \in P_0 \cup A_{\vec{F}}$, if $\exists t \in p^\bullet$, $^\bullet t \cap P_R \subseteq R_{\bar{S}}$, then $t \in T_{\vec{F}}$ and $^\bullet t \cap P_A \subseteq A_{\vec{F}}$.*

From Definition 4.6, we have that each sink transition in a DP-net belongs to $T_{\vec{F}}$ and the activity pre-places of any sink transition belong to $A_{\vec{F}}$.

Definition 4.7 *$T_{\overleftarrow{F}}$ and $A_{\overleftarrow{F}}$ are recursively defined as: $\forall p \in P_0 \cup A_{\overleftarrow{F}}$, if $\exists t \in p^\bullet$, $^\bullet t \cap P_R \subseteq R_{\bar{S}}$, then $t \in T_{\overleftarrow{F}}$ and $t^\bullet \cap P_A \subseteq A_{\overleftarrow{F}}$. The set of critical activity places is $A_K = P_A \setminus (A_{\vec{F}} \cup A_{\overleftarrow{F}})$ and the set of critical transitions is $T_K = T \setminus (T_{\vec{F}} \cup T_{\overleftarrow{F}})$.*

An illustrative example is given in Section 4.5. The following property is derived from the above definitions.

Property 4.2 *$\forall t \in T_{\vec{F}}$ (respectively, $\forall t \in T_{\overleftarrow{F}}$), $\forall p \in t^\bullet \cap P_A$, we have $p \in A_{\vec{F}}$ (respectively, $p \in A_{\overleftarrow{F}}$) and the operation represented by p does not require any shared resources, i.e., $U_p|_{R_S} = 0$ where $U_p|_{R_S}$ denotes a resource requirement vector U_p restricted to the places in R_S.*

Property 4.2 means that any output activity place of a free transition is a free activity place. Note that all the last activity places in a DP-net belong to $A_{\vec{F}}$, i.e.,

$\forall x \in \mathbb{N}_J$, $^{\bullet\bullet}p_0^x \cap P_A \subseteq A_{\vec{F}}$. The reason is that, for any last activity place p, we have that $\forall t \in p^\bullet$, t is a sink transition of the same subnet. By Definition 3.1, $^\bullet t \cap P_R = \emptyset \subseteq R_{\bar{S}}$. Thus, $t \in T_{\vec{F}}$ and $p \subseteq A_{\vec{F}}$ according to Definition 4.6. It also implies that A_F is a proper subset of P_A, i.e., $A_F \subset P_A$. Algorithm 4.1 is used to generate T_F, A_F, and A_K for a DP-net.

Algorithm 4.1

```
Generation of critical and free activity places.
Input: A DP-net N = (P₀ ∪ Pᵣ ∪ Pₐ, T, F, W) with Rₛ and Rₛ̄.
Output: Tₚ, Aₚ, and Aₖ.
Compute_critical_activity_places {
        A⃗ₚ := ∅; Aₚ̄ := ∅; T⃗ₚ := ∅; Tₚ̄ := ∅;
        for each p ∈ P₀
        {
              RIGHT(p, A⃗ₚ, T⃗ₚ);
              LEFT(p, Aₚ̄, Tₚ̄);
        }
        Tₚ := T⃗ₚ ∪ Tₚ̄; Aₚ := A⃗ₚ ∪ Aₚ̄; Aₖ := Pₐ \ Aₚ;
        Output Tₚ, Aₚ, and Aₖ;
}
Function RIGHT(p, A⃗ₚ, T⃗ₚ) {
        for each p' ∈ ••p ∩ Pₐ
              if •(p'•) ∩ Pᵣ ⊆ Rₛ̄
              {
                    A⃗ₚ := A⃗ₚ ∪ {p'};
                    T⃗ₚ := T⃗ₚ ∪ p'•;
                    RIGHT(p', A⃗ₚ, T⃗ₚ);
              }
}
Function LEFT(p, Aₚ̄, Tₚ̄) {
        for each p' ∈ p•• ∩ Pₐ
              if ••p' ∩ Pᵣ ⊆ Rₛ̄
              {
                    Aₚ̄ := Aₚ̄ ∪ {p'};
                    Tₚ̄ := Tₚ̄ ∪ •p';
                    LEFT(p', Aₚ̄, Tₚ̄);
              }
}
```

The computational complexity of Algorithm 4.1 is briefly analyzed as follows. The algorithm searches free activity places among P_A in two opposite directions from idle places by using two recursions. At each invocation of the function RIGHT or LEFT in the recursions, a place in $A_{\vec{F}}$ or $A_{\bar{F}}$ is identified. When the recursions terminate, all places in $A_{\vec{F}}$ and $A_{\bar{F}}$ are found. The number of invocations of the functions RIGHT and LEFT in the algorithm is $|A_{\vec{F}}| + |A_{\bar{F}}|$. In the worst case where all places in P_A are free activity ones, the number of places searched by the algorithm is linear with $|P_A|$.

4.3 Properties of DP-Nets

This section presents some properties of DP-nets, which are to be used in the supervisor synthesis based on critical activity places.

Definition 4.8 *In a DP-net N, a string of $p_1, t_1, p_2, t_2, ..., p_n, t_n$ with $n \geq 1$ is called a path of $A^x_{\overrightarrow{F}}$ with $x \in \mathbb{N}_J$ if $p_1, p_2, ..., p_n \in A^x_{\overrightarrow{F}}$ and $(p_1, t_1), (t_1, p_2), ..., (p_n, t_n) \in F^x$. The path is said to be a full path of $A^x_{\overrightarrow{F}}$ if $^{\bullet\bullet}p_1 \cap A^x_{\overrightarrow{F}} = \emptyset$ and $t^{\bullet}_n = p^x_0$.*

Definition 4.9 *In a DP-net N, a string of $p_1, t_1, p_2, t_2, ..., p_n, t_n$ with $n \geq 1$ is called a path of $A^x_{\overleftarrow{F}}$ with $x \in \mathbb{N}_J$ if $p_1, p_2, ..., p_n \in A^x_{\overleftarrow{F}}$ and $(t_n, p_n), (p_n, t_{n-1}), ..., (t_1, p_1) \in F^x$. The path is said to be a full path of $A^x_{\overleftarrow{F}}$ if $^{\bullet}t_n = p^x_0$ and $p^{\bullet\bullet}_1 \cap A^x_{\overleftarrow{F}} = \emptyset$.*

Definition 4.10 *Given two places $p, p' \in A_{\overrightarrow{F}}$ (respectively, $A_{\overleftarrow{F}}$), if there exists a path, $..., p, ..., p', ...,$ in $A^x_{\overrightarrow{F}}$ (respectively, $A^x_{\overleftarrow{F}}$) with $x \in \mathbb{N}_J$, p' is called a succeeding place of p in $A_{\overrightarrow{F}}$ (respectively, $A_{\overleftarrow{F}}$), denoted as $p' \in \overrightarrow{Suc}(p)$ (respectively, $p' \in \overleftarrow{Suc}(p)$).*

The following properties are derived from Definitions (4.8)–(4.10).

Property 4.3 *For a place $p \in A_{\overrightarrow{F}}$ (respectively, $p \in A_{\overleftarrow{F}}$), there exists a full path of $A^x_{\overrightarrow{F}}$ (respectively, $A^x_{\overleftarrow{F}}$) with $x \in \mathbb{N}_J$ that contains p.*

Property 4.4 *Let p be a place in $A^x_{\overrightarrow{F}}$ with $x \in \mathbb{N}_J$. If p is not the first place of any full path of $A^x_{\overrightarrow{F}}$, then p does not require any shared resource, i.e., $U_p|_{R_S} = 0$.*

Property 4.5 *Let t be a transition of a path of $A^x_{\overrightarrow{F}}$. The firing of t does not need any shared resource, i.e., $^{\bullet}t \cap R_S = \emptyset$.*

Extended from the definition of MTSI in Li et al. (2008a), the set of zone separation instances is defined as follows.

Definition 4.11 *Let $G(N, M_0)$ be the reachability graph of a DP-net $N = (P_0 \cup P_R \cup P_A, T, F, W)$ with an initial marking M_0. Let $\mathcal{M}_L$ and $\mathcal{M}_D$ be the sets of markings in the live zone and deadlock zone of $G(N, M_0)$, respectively. The set $Z = \{(M, t, M')| M[t\rangle M' \wedge t \in T \wedge M \in \mathcal{M}_L \wedge M' \in \mathcal{M}_D\}$ is called the set of zone separation instances of the net.*

Next, we prove that the transition in any zone separation instance is a critical transition.

Theorem 4.1 $\forall (M, t, M') \in Z, t \in T_K$.

Proof: We first show that $\forall (M, t, M') \in Z$, $t \notin T_{\overleftarrow{F}}$. Suppose that there exists $(M_1, t_1, M_2) \in Z$ such that $t_1 \in T_{\overleftarrow{F}}$. By Property 4.2, we have $\exists p_1 \in P$, $t_1{}^{\bullet} \cap P_A = \{p_1\} \subseteq A_{\overleftarrow{F}}$ and p_1 does not use any shared resource, i.e., $U_{p_1}|_{R_S} = \mathbf{0}$. So, the firing of t_1, which is a pre-transition of p_1, does not need any shared resource. In a DP-net, deadlocks and livelocks are caused by competitions for shared resources. Thus, if M_1 is not a deadlock, a livelock, or a marking that inevitably evolves to a deadlock or a livelock, i.e., $M_1 \notin \mathcal{M}_D$, then its immediate successor resulting from the firing of t_1 is not a deadlock, livelock, or bad one doomed to a deadlock or livelock. That is to say, if $M_1 \in \mathcal{M}_L$ and $M_1[t_1\rangle M_2$, then $M_2 \in \mathcal{M}_L$. Thus, $(M_1, t_1, M_2) \notin Z$. It contradicts the assumption. Therefore, $\forall (M, t, M') \in Z$, $t \notin T_{\overleftarrow{F}}$.

On the other hand, we can similarly prove that $\forall (M, t, M') \in Z$, $t \notin T_{\overrightarrow{F}}$. Thus, $\forall (M, t, M') \in Z$, $t \notin T_{\overleftarrow{F}} \cup T_{\overrightarrow{F}} = T_F$. In addition, $T_F \cup T_K = T$ and $T_F \cap T_K = \emptyset$. Therefore, $\forall (M, t, M') \in Z$, $t \in T_K$ holds. ∎

In the sequel, $T_{\overrightarrow{F}}^*$ (respectively, $T_{\overleftarrow{F}}^*$) is used to denote the set of all finite strings of elements in $T_{\overrightarrow{F}}$ (respectively, $T_{\overleftarrow{F}}$) including the empty string.

Lemma 4.1 $\forall M \in R(N, M_0)$ of a DP-net N, if $p \in A_{\overrightarrow{F}}$ is a marked place at M, then p can be emptied by firing a string of transitions in $T_{\overrightarrow{F}}^*$ without affecting the tokens in $P_A \backslash (p \cup \overrightarrow{Suc}(p))$.

Proof: We prove it by induction. $\forall M \in R(N, M_0)$, if $p \in A_{\overrightarrow{F}}$ and $M(p) > 0$, then $p \in A_{\overrightarrow{F}}^x$ with $x \in \mathbb{N}_J$ and p is in a full path of $A_{\overrightarrow{F}}^x$ according to Property 4.3. If p is the last place in the full path of $A_{\overrightarrow{F}}^x$, then p can be emptied by firing any transition in $p^{\bullet} \subseteq T_{\overrightarrow{F}}$ since $\forall t \in p^{\bullet}$, t is a sink transition and the firing of any sink transition of a DP-net does not need any resources. In addition, the firing of t only changes the token distribution in p, p_0^x, and some resource places in P_R, which satisfies $(p \cup p_0^x \cup P_R) \cap (P_A \backslash (p \cup \overrightarrow{Suc}(p))) = \emptyset$.

Suppose that if p is the nth place from the end of the full path, p can be emptied by firing a string of transitions in $T_{\overrightarrow{F}}^*$ without affecting the tokens in $P_A \backslash (p \cup \overrightarrow{Suc}(p))$. Now suppose that p is the $(n + 1)$th place and p' be the nth place from the end of the full path. It means that p' is the succeeding place of p in the path. According to the assumption, if p' is marked, it can be emptied by firing a sequence of transitions in $T_{\overrightarrow{F}}^*$ without affecting the token distribution in $P_A \backslash (p' \cup \overrightarrow{Suc}(p'))$. Since p is the previous activity place of p' in the full path, we have $p' \cup \overrightarrow{Suc}(p') \subseteq p \cup \overrightarrow{Suc}(p)$. Thus, p' can be emptied without affecting the token distribution in $P_A \backslash (p \cup \overrightarrow{Suc}(p))$. In addition, by Property 4.5, the firing of t, which is the transition between p and p', does not need any shared resources. So, if $M(p) > 0$, one token can be moved from p to p' by firing t whenever p' has available space to accept this token. Next, the token in p'

can be removed without changing the tokens in $P_A\backslash(p \cup \overrightarrow{Suc}(p))$ and the resources used by p' are also released. Then, t can fire again to move another token from p to p'. The above process can be performed many times until there is no token in p. In addition, $t \in T_{\overleftarrow{F}}$ and the firing of t only affects the tokens of p, p', and some places in P_R with $(p \cup p' \cup P_R) \cap (P_A\backslash(p \cup \overrightarrow{Suc}(p))) = \emptyset$. Therefore, for $p \in A_{\overleftarrow{F}}$, if $M(p) > 0$, then p can be emptied by firing a string of transitions in $T_{\overleftarrow{F}}^*$ without affecting the tokens in $P_A\backslash(p \cup \overrightarrow{Suc}(p))$. ∎

Lemma 4.2 $\forall M \in R(N, M_0)$ *of a DP-net N, if $p \in A_{\overrightarrow{F}}$ is a marked place at M, then p can be emptied by conversely firing a string of transitions in $T_{\overrightarrow{F}}^*$ without affecting the tokens in $P_A\backslash(p \cup \overleftarrow{Suc}(p))$.*

Proof: It can be similarly proven as Lemma 4.1. ∎

Theorem 4.2 $\forall M \in \mathcal{M}_D$, *there exist $\sigma_1 \in T_{\overrightarrow{F}}^*$, $\sigma_2 \in T_{\overleftarrow{F}}^*$, and M' with $M'(A_K) = M(A_K)$ and $M'(A_F) = \mathbf{0}$ such that $M[\overrightarrow{\sigma_1} \cdot \overleftarrow{\sigma_2}\rangle M'$ and $M' \in \mathcal{M}_D$.*

Proof: The proof is divided into two parts.

(i) We first prove that $\forall M \in \mathcal{M}_D$, all places in $A_{\overrightarrow{F}}$ can be emptied by firing a sequence of transitions in $T_{\overrightarrow{F}}^*$ without affecting the marking in $P_A\backslash A_{\overrightarrow{F}}$. $\forall x \in \mathbb{N}_J$, for any full path, $p_1, t_1, p_2, t_2, \ldots, p_n, t_n$, of $A_{\overrightarrow{F}}^x$, if the first place p_1 of the path is not empty, i.e., $M(p_1) > 0$, then p_1 can be emptied by firing a transition sequence of $T_{\overrightarrow{F}}^*$ without affecting the marking of $P_A\backslash(p_1 \cup \overrightarrow{Suc}(p_1))$ according to Lemma 4.1. That is to say, for P_A, this process can only change the token distribution in p_1 and its succeeding places in $A_{\overrightarrow{F}}^x$. Then, if the second place p_2 of the path is marked, it can be emptied by firing a transition sequence of $T_{\overrightarrow{F}}^*$ without affecting the marking of $P_A\backslash(p_2 \cup \overrightarrow{Suc}(p_2))$ including the emptied place p_1. Similarly, by firing a transition sequence of $T_{\overrightarrow{F}}^*$, we can empty the rest of the places in the full path one by one until the last place. It changes the marking of the current place and its succeeding places of the path but does not add any tokens to its preceding places which have already been emptied. In other words, all places of the full path can be emptied by firing a transition sequence of $T_{\overrightarrow{F}}^*$ without changing the marking of $P_A\backslash A_{\overrightarrow{F}}^x$. Since $A_{\overrightarrow{F}}^x \subseteq A_{\overrightarrow{F}}$, we have $P_A\backslash A_{\overrightarrow{F}} \subseteq P_A\backslash A_{\overrightarrow{F}}^x$. So, the above process does not affect the marking of $P_A\backslash A_{\overrightarrow{F}}$. Thus, $\forall M \in \mathcal{M}_D$, $\forall x \in \mathbb{N}_J$, all places of any full path of $A_{\overrightarrow{F}}^x$ can be emptied by firing a transition sequence of $T_{\overrightarrow{F}}^*$ without affecting the marking of $P_A\backslash A_{\overrightarrow{F}}$. Also, by Property 4.3, $\forall p \in A_{\overrightarrow{F}}$, p belongs to a full path of $A_{\overrightarrow{F}}^x$ with $x \in \mathbb{N}_J$. Therefore, $\forall M \in \mathcal{M}_D$, all places in $A_{\overrightarrow{F}}$ can be emptied by firing a transition sequence of $T_{\overrightarrow{F}}^*$ without changing the marking of $P_A\backslash A_{\overrightarrow{F}}$.

Figure 4.1 Set relations between P_A, A_K, and A_F.

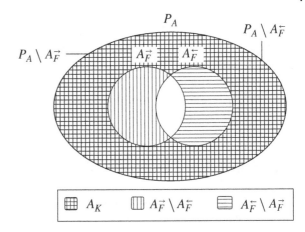

Let σ_1 be a transition sequence of $T^*_{\overrightarrow{F}}$ and M'' be a marking obtained by firing σ_1 such that $M''(P_A \backslash A_{\overrightarrow{F}}) = M(P_A \backslash A_{\overrightarrow{F}})$ and $M''(A_{\overrightarrow{F}}) = \mathbf{0}$. Then we have $M[\overrightarrow{\sigma_1}\rangle M''$. Since $M \in \mathcal{M}_D$ and M'' is reachable from M by firing σ_1, M'' belongs to the deadlock zone.

(ii) Similarly, we can prove that for a marking $M'' \in \mathcal{M}_D$, all places in $A_{\overleftarrow{F}}$ can be emptied by conversely firing a transition sequence of $T^*_{\overleftarrow{F}}$ without changing the marking of $P_A \backslash A_{\overleftarrow{F}}$. Let σ_2 be such a transition sequence of $T^*_{\overleftarrow{F}}$ and M' be the resultant marking such that $M'(P_A \backslash A_{\overleftarrow{F}}) = M''(P_A \backslash A_{\overleftarrow{F}})$ and $M'(A_{\overleftarrow{F}}) = \mathbf{0}$. Then we have $M''[\overleftarrow{\sigma_2}\rangle M'$.

Therefore, $M[\overrightarrow{\sigma_1} \cdot \overleftarrow{\sigma_2}\rangle M'$. By substituting M'' into M', we have that $M'((P_A \backslash A_{\overrightarrow{F}}) \cap (P_A \backslash A_{\overleftarrow{F}})) = M((P_A \backslash A_{\overrightarrow{F}}) \cap (P_A \backslash A_{\overleftarrow{F}}))$, $M'(A_{\overleftarrow{F}}) = \mathbf{0}$, and $M'(A_{\overrightarrow{F}} \backslash A_{\overleftarrow{F}}) = \mathbf{0}$. Since $A_F = A_{\overleftarrow{F}} \cup A_{\overrightarrow{F}}$ and $A_K = P_A \backslash A_F$, then $A_F = A_{\overleftarrow{F}} \cup (A_{\overrightarrow{F}} \backslash A_{\overleftarrow{F}})$ and $A_K = (P_A \backslash A_{\overrightarrow{F}}) \cap (P_A \backslash A_{\overleftarrow{F}})$ according to the set relations shown in Figure 4.1. Thus, we have $M'(A_K) = M(A_K)$ and $M'(A_F) = \mathbf{0}$. Suppose that $\sigma_2 = t_1 t_2 \dots t_n$ and $M_1, M_2, \dots, M_{n-1}$ are the markings such that $M''[\overleftarrow{t_1}\rangle M_1 [\overleftarrow{t_2}\rangle M_2 \dots M_{n-1}[\overleftarrow{t_n}\rangle M'$ with $t_1, t_2, \dots, t_n \in T_{\overleftarrow{F}} \subseteq T_F \neq T_K$. By Theorem 4.1, since $t_1 \notin T_K$, we have $(M_1, t_1, M'') \notin Z$. Thus, $M'' \in \mathcal{M}_D$ implies $M_1 \notin \mathcal{M}_L$. Similarly, $M_2, \dots, M_{n-1}$, and M' are not in $\mathcal{M}_L$ either. Therefore, $M' \in \mathcal{M}_D$ holds. ∎

Corollary 4.1 $\forall M \in \mathcal{M}_L$, there exist $\sigma_1 \in T^*_{\overrightarrow{F}}$, $\sigma_2 \in T^*_{\overleftarrow{F}}$, and M' with $M'(A_K) = M(A_K)$ and $M'(A_F) = \mathbf{0}$ such that $M[\overrightarrow{\sigma_1} \cdot \overleftarrow{\sigma_2}\rangle M'$ and $M' \in \mathcal{M}_L$.

Theorem 4.3 *In a DP-net, at least two critical activity places are marked at any FBM.*

Proof: It is proven by contradiction. Suppose that at most one critical activity place is marked at any FBM M. According to Theorem 4.2, since $M \in \mathcal{M}_F \subseteq \mathcal{M}_D$, *all tokens in A_F at M can be removed by firing some transitions without affecting the tokens in A_K. In addition, the resultant marking, denoted as M', belongs to the deadlock zone, i.e., $M' \in \mathcal{M}_D$. According to the assumption, there is at most one activity place that is marked at M'. Then, it is impossible for M' to have or will have a deadlock or a livelock if at most one activity is being executed in the system, because it implies that at most one activity place is marked and a circular wait is not possible for M'. So, $M' \notin \mathcal{M}_D$. It is a contradiction.* ∎

4.4   Supervisor Design with Critical Activity Places

In this section, we show that place invariants on critical activity places A_K are sufficient for obtaining a Petri net supervisor for a given DP-net. First, we consider the place invariant design method where the coefficients of the designed place invariants are in $\{0, 1\}$, such as the method in Uzam and Zhou (2007). In order to forbid an FBM $M' \in \mathcal{M}_F$, the marking of critical activity places can be used in a place invariant, $\sum_{p_i \in A_K} l(i) \cdot M(p_i) + M(p_c) = \beta \leqslant \sum_{p_i \in A_K} M'(p_i) - 1$ where p_c is the designed control place related to the place invariant, β is the constant sum of the place invariant, and $l(i) = 1$ if $M'(p_i) > 0$, otherwise $l(i) = 0$. We call it a place invariant on A_K since only places in A_K are considered in constructing the place invariant. This place invariant forbids FBM M'. When all FBMs in $\mathcal{M}_F$ are forbidden by such designed place invariants on A_K, the system is kept running in the live zone of the reachability graph. Then, a Petri net supervisor can be computed by the following equations.

$$[N_d] = -l \cdot [N_{A_K}] \tag{4.3}$$

$$M_0(P_c) = \beta - l \cdot M_0(A_K) \tag{4.4}$$

where P_c is the set of control places that correspond to the designed place invariants on A_K, $[N_d]$ is the incidence matrix of P_c and T, $[N_{A_K}]$ is the incidence matrix of A_K and T, and l is the coefficient matrix of designed place invariants. Please note that such a supervisor is an ordinary net. The reason is as follows. $[N_{A_K}]$ consists of some rows in $[N_{P_A}]$ and there exist three cases for a column in $[N_{A_K}]$ according to Property 4.1, i.e., i) one entry is 1 and the rest are 0, ii) one entry is −1 and the rest are 0, and iii) two entries are 1 and −1 respectively and the rest are 0. In addition, the entries of the coefficients matrix l are in $\{0, 1\}$. Thus, by Equation (4.3), each entry in $[N_d]$ is in $\{0, 1, -1\}$, which means that the weight of any arc between control places and transitions equals one.

Then, we show that the design of place invariants on A_K is also sufficient for the supervisor synthesis where the coefficients of the place invariants are greater than one, which implies that the obtained supervisor is a generalized net.

Definition 4.12 *Let M and M' be two markings in $R(N, M_0)$. M K-covers M' (respectively, M' is K-covered by M) if $\forall p \in A_K$, $M(p) \geqslant M'(p)$, which is denoted as $M \geqslant_K M'$ (respectively, $M' \leqslant_K M$).*

Corollary 4.2 *Let M and M' be two markings in $R(N, M_0)$ with $M \geqslant_K M'$. If M' is forbidden by a place invariant, then M is also forbidden by the place invariant. If M is permitted by a place invariant, then M' is also permitted by it.*

Theorem 4.4 *For an FBM $M' \in \mathcal{M}_F$, the constraint*

$$\sum_{p_i \in A_K} l(i) \cdot (M(p_i) - M'(p_i)) \leqslant -1, \quad \forall M \in \mathcal{M}_L \tag{4.5}$$

forbids M' and permits all legal markings in $\mathcal{M}_L$.

Proof: As shown in Section 3.3.2, the following place invariant is used to construct a monitor in a supervisor.

$$\sum_{p_i \in P_A} l(i) \cdot M(p_i) + M(p_c) = \beta. \tag{4.6}$$

The place invariant can forbid a marking $M \in R(N, M_0)$ if

$$\sum_{p_i \in P_A} l(i) \cdot M(p_i) \geqslant \beta + 1. \tag{4.7}$$

Such a place invariant is called a place invariant on P_A since only activity places of the plant nets are considered to construct the place invariant. By Theorem 4.2, $\forall M' \in \mathcal{M}_F \subseteq \mathcal{M}_D$, $\exists M'' \in \mathcal{M}_D$ such that $M''(A_K) = M'(A_K)$ and $M''(A_F) = \mathbf{0}$. In addition, we have $M' \geqslant_K M''$ according to Definition 4.12. Thus, by Corollary 4.2, if M'' is forbidden by a place invariant, i.e., $\sum_{p_i \in P_A} l(i) \cdot M''(p_i) \geqslant \beta + 1$, M' is also forbidden by the place invariant. On the other hand, $M''(A_F) = \mathbf{0}$ implies that the weighted sum of the tokens in A_F is zero, i.e., $\sum_{p_i \in A_F} l(i) \cdot M''(p_i) = 0$. Therefore, there is no need for the place invariant to impose constraints on A_F to forbid M''. In other words, it is sufficient to forbid M'' and M' by the place invariant on A_K such that

$$\sum_{p_i \in A_K} l(i) \cdot M'(p_i) \geqslant \beta + 1. \tag{4.8}$$

In addition, the place invariant permits all legal markings if

$$\sum_{p_i \in A_K} l(i) \cdot M(p_i) \leqslant \beta, \quad \forall M \in \mathcal{M}_L. \tag{4.9}$$

By substituting (4.9) into (4.8), the constraint to forbid M' and permit all legal markings becomes

$$\sum_{p_i \in A_K} l(i) \cdot (M(p_i) - M'(p_i)) \leqslant -1, \quad \forall M \in \mathcal{M}_L. \tag{4.10}$$

∎

Theorem 4.4 indicates that the critical activity places A_K are sufficient for constructing a place invariant to forbid an FBM and permit all legal markings. Next, we show that only two small sets of markings in $\mathcal{M}_F$ and $\mathcal{M}_L$ need to be considered in the supervisor synthesis via place invariants on A_K.

Definition 4.13 $\mathcal{M}_F^K$ *is called a minimal covered set of* $\mathcal{M}_F$ *with respect to K-cover if*

1) $\mathcal{M}_F^K \subseteq \mathcal{M}_F$;
2) $\forall M \in \mathcal{M}_F, \exists M' \in \mathcal{M}_F^K$ *such that* $M \geqslant_K M'$; *and*
3) $\forall M \in \mathcal{M}_F^K, \nexists M'' \in \mathcal{M}_F^K$ *such that* $M \geqslant_K M''$ *and* $M \neq M''$.

Corollary 4.3 *If all markings in* $\mathcal{M}_F^K$ *are forbidden by some place invariants on* A_K, *then all markings in* $\mathcal{M}_F$ *are also forbidden by them.*

Proof: *It follows immediately from Definition 4.13 and Corollary 4.2.* ∎

Corollary 4.3 indicates that only a minimal covered set of $\mathcal{M}_F$ with respect to K-cover, i.e., $\mathcal{M}_F^K$, needs to be considered in the design of place invariants on A_K to prevent the system from entering the deadlock zone.

Definition 4.14 $\mathcal{M}_L^K$ *is called a minimal covering set of* $\mathcal{M}_L$ *with respect to K-cover if*

1) $\mathcal{M}_L^K \subseteq \mathcal{M}_L$;
2) $\forall M \in \mathcal{M}_L, \exists M' \in \mathcal{M}_L^K$ *subject to* $M' \geqslant_K M$; *and*
3) $\forall M \in \mathcal{M}_L^K, \nexists M'' \in \mathcal{M}_L^K$ *subject to* $M'' \geqslant_K M$ *and* $M \neq M''$.

Corollary 4.4 *Given an FBM* $M \in \mathcal{M}_F^K$. *If a place invariant on* A_K *forbids M and permits all markings in* $\mathcal{M}_L^K$, *then it also permits all markings in* $\mathcal{M}_L$.

Proof: *It follows immediately from Definition 4.14 and Corollary 4.2.* ∎

Corollary 4.4 indicates that for an FBM, only a minimal covering set of $\mathcal{M}_L$ with respect to K-cover, i.e., $\mathcal{M}_L^K$, needs to be considered in the design of a place invariant on A_K to forbid the FBM and permit all legal markings. Therefore, for an FBM $M' \in \mathcal{M}_F^K \subseteq \mathcal{M}_F$, (4.5) can be simplified as follows:

$$\sum_{p_i \in A_K} l(i) \cdot (M(p_i) - M'(p_i)) \leqslant -1, \quad \forall M \in \mathcal{M}_L^K. \tag{4.11}$$

To forbid all markings in $\mathcal{M}_F^K$, we use j ($j = 1, 2, ..., |\mathcal{M}_F^K|$) candidate place invariants on A_K with the coefficients $l_j(i)$, each of which forbids $M_j \in \mathcal{M}_F^K$ and permits all legal markings in $\mathcal{M}_L^K$. Thus, we have

$$\sum_{p_i \in A_K} l_j(i) \cdot (M_m(p_i) - M_j(p_i)) \leqslant -1,$$

$$\forall M_j \in \mathcal{M}_F^K \text{ and } \forall M_m \in \mathcal{M}_L^K. \tag{4.12}$$

Note that each place invariant on A_K may forbid more than one marking in $\mathcal{M}_F^K$. Given a place invariant I_j that forbids $M_j \in \mathcal{M}_F^K$, $\exists M_k \in \mathcal{M}_F^K$ with $k \neq j$, M_k is also forbidden by I_j if:

$$\sum_{p_i \in A_K} l_j(i) \cdot (M_k(p_i) - M_j(p_i)) \geqslant -\Gamma \cdot (1 - f_{j,k}),$$

$$\forall M_k \in \mathcal{M}_F^K \text{ and } k \neq j \tag{4.13}$$

where Γ is a big enough positive integer and $f_{j,k} \in \{0, 1\}$ such that $f_{j,k} = 1$ indicates that I_j designed to forbid M_j also forbids M_k and $f_{j,k} = 0$ indicates that I_j does not forbid M_k.

In addition, a variable q_j ($j \in \mathbb{N}_F^K = \{i | M_i \in \mathcal{M}_F^K\}$) for I_j is introduced to indicate whether I_j is picked to design a control place or not. $q_j = 1$ implies that I_j is picked and $q_j = 0$ implies that it is not. The following constraint represents that only when I_j is picked, can it forbid M_j and the FBMs that satisfy (4.13).

$$f_{j,k} \leqslant q_j, \quad \forall j, k \in \mathbb{N}_F^K \text{ and } k \neq j. \tag{4.14}$$

To ensure that all markings in $\mathcal{M}_F^K$ are forbidden by the selected place invariants, $f_{k,j}$ and q_j should satisfy the following constraint:

$$q_j + \sum_{k \in \mathbb{N}_F^K, k \neq j} f_{k,j} \geqslant 1, \quad \forall j \in \mathbb{N}_F^K. \tag{4.15}$$

By combining the above constraints with an objective to minimize the number of selected place invariants on A_K, which implies the fewest control places in the supervisor, we have the following ILP problem.

$$\min \sum_{j \in \mathbb{N}_F^K} q_j \tag{4.16}$$

subject to

$$\sum_{p_i \in A_K} l_j(i) \cdot (M_m(p_i) - M_j(p_i)) \leqslant -1,$$

$$\forall M_j \in \mathcal{M}_F^K \text{ and } \forall M_m \in \mathcal{M}_L^K$$

$$\sum_{p_i \in A_K} l_j(i) \cdot (M_k(p_i) - M_j(p_i)) \geqslant -\Gamma \cdot (1 - f_{j,k}),$$

$$\forall M_k \in \mathcal{M}_F^K \text{ and } k \neq j$$

$$f_{j,k} \leqslant q_j, \quad \forall j, k \in \mathbb{N}_F^K \text{ and } k \neq j$$

$$q_j + \sum_{k \in \mathbb{N}_F^K, k \neq j} f_{k,j} \geq 1, \quad \forall j \in \mathbb{N}_F^K$$

$$l_j(i) \in \{0, 1, 2, \ldots\}, \quad \forall p_i \in A_K \text{ and } \forall j \in \mathbb{N}_F^K$$

$$f_{j,k} \in \{0, 1\}, \quad \forall j, k \in \mathbb{N}_F^K \text{ and } k \neq j$$

$$q_j \in \{0, 1\}, \quad \forall j \in \mathbb{N}_F^K$$

Similarly to Chen and Li (2011, Theorem 1), if the ILP in (4.16) has an optimal solution, the designed place invariants on A_K can keep the system from entering the deadlock zone and permit all legal markings in the live zone. Then, by using (4.3) and (4.4), we can construct an optimal supervisor for the original net. In addition, the number of inequalities in (4.16) is $|\mathcal{M}_F^K| \cdot (|\mathcal{M}_L^K| + 2|\mathcal{M}_F^K| - 1)$, which is fewer than that of the ILP in Chen and Li (2011) because only critical activity places are considered in the constructions of place invariants.

4.5 An Illustrative Example

As an example, we consider a conjunctive/disjunctive resource allocation system that is modified from Park and Reveliotis (2001), which allows for multiple resource acquisitions and flexible routings. It has two product types P1 and P2 and six resources types R_1–R_6 in which R_2 has two resource units, R_3 has three resource units, and the rest has one resource unit each. The processing routes of P1 (respectively, P2) are given by the set of partially ordered job stages $\{p_2, p_3, p_4, p_5, p_6, p_7, p_8\}$ (respectively, $\{p_{10}, p_{11}, p_{12}, p_{13}\}$). The resource requirements of operations are as follows: $U_{p_2} = (1, 0, 0, 0, 0, 0)$, $U_{p_3} = (0, 0, 0, 1, 0, 0)$, $U_{p_4} = (0, 1, 1, 0, 0, 0)$, $U_{p_5} = (0, 1, 0, 0, 0, 0)$, $U_{p_6} = (0, 2, 0, 0, 0, 0)$, $U_{p_7} = (0, 2, 1, 0, 0, 0)$, $U_{p_8} = (0, 3, 0, 0, 0, 0)$, $U_{p_{10}} = (0, 0, 1, 0, 0, 0)$, $U_{p_{11}} = (0, 0, 0, 0, 1, 0)$, $U_{p_{12}} = (0, 3, 0, 0, 0, 0)$, and $U_{p_{13}} = (0, 0, 0, 0, 0, 1)$.

Its DP-net model is shown in Figure 4.2 that has 19 places and 15 transitions. It is a generalized Petri net. The descriptions of places are shown in Table 4.1. The meanings of transitions are omitted since they similarly denote the start and/or end of operations. The reachability graph of the net has 620 markings, of which 508 are legal markings and 72 are FBMs. By using Algorithm 4.1, we have $T_{\overline{F}}\{t_1, t_2\}$, $T_{\overline{F}}\{t_8, t_{10}, t_{14}, t_{15}\}$, $T_K\{t_3 - t_7, t_9, t_{11} - t_{13}\}$, $A_F = \{p_2, p_3\}$, $A_{\overline{F}} = \{p_7, p_8, p_{12}, p_{13}\}$, and $A_K = \{p_4, p_5, p_6, p_{10}, p_{11}\}$. By using the K-cover method, $\mathcal{M}_L^K$ and $\mathcal{M}_F^K$ have nine and five markings, respectively. Then, place invariants on A_K are constructed by using the ILP in (4.16) to design a supervisor that forbids all markings in $\mathcal{M}_F^K$ and permits all markings in $\mathcal{M}_L^K$. The ILP has 90 constraints, 30 fewer constraints than that of Chen and Li (2011) since

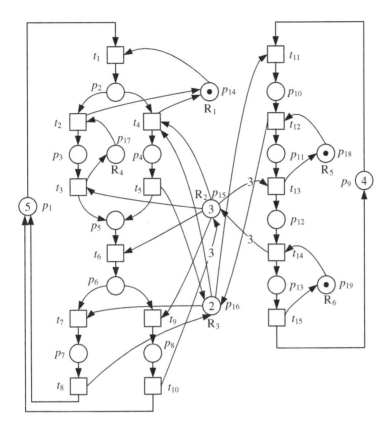

Figure 4.2 DP-net of an RAS.

fewer places are needed in the place invariant construction. As a result, the obtained optimal supervisor has one control place with $^{\bullet}p_c = \{3t_7, 3t_9, t_{13}\}$, $p_c^{\bullet} = \{3t_3, 3t_4, t_{11}\}$, and $M_0(p_c) = 8$. Figure 4.3 shows the derived supervisor. After adding it to the original net, the controlled net becomes live and can reach all 508 legal markings of the original net.

Note that Cordone and Piroddi (2011) mention that the activity places that do not use critical resources are useless in the place invariant design. But the definition of critical resources is not given. Furthermore, some activity places that do not use shared resources, e.g., p_{11} in Figure 4.2, are useful in the place invariant design. In Uzam (2004) and Uzam and Zhou (2006), a net reduction method for the place invariant design is given based on net fusion and elimination rules. The method presented in this chapter can further remove the places that are useless in the place invariant design, e.g., p_2 and p_3 in Figure 4.2, which are used in the method of Uzam (2004) and Uzam and Zhou (2006).

Table 4.1 Descriptions of a system model.

p_1	Raw parts of product P1 are available
p_2	Product P1 is being processed by oneR$_1$
p_3	Product P1 is being processed by oneR$_4$
p_4	Product P1 is being processed by oneR$_2$ and oneR$_3$
p_5	Product P1 is being processed by oneR$_2$
p_6	Product P1 is being processed by twoR$_2$
p_7	Product P1 is being processed by twoR$_2$ and oneR$_3$
p_8	Product P1 is being processed by threeR$_2$
p_9	Raw parts of product P2 are available
p_{10}	Product P2 is being processed by oneR$_3$
p_{11}	Product P2 is being processed by oneR$_5$
p_{12}	Product P2 is being processed by threeR$_2$
p_{13}	Product P2 is being processed by oneR$_6$
p_{14}	R$_1$ is available
p_{15}	R$_2$ is available
p_{16}	R$_3$ is available
p_{17}	R$_4$ is available
p_{18}	R$_5$ is available
p_{19}	R$_6$ is available

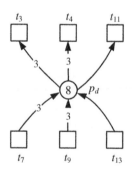

Figure 4.3 Supervisor for the DP-net shown in Figure 4.2.

4.6 Concluding Remarks

For practitioners who are working in the area of supervisory control for RASs, deadlock prevention is an important issue since deadlocks may lead to catastrophic results in RASs. Based on Petri nets, deadlock prevention methods always aim to design a supervisor to ensure that deadlocks never occur. Existing reachability analysis methods for deadlock prevention require that all activity places are considered in the place invariant design. However, some activity places may be useless and thus result in redundant constraints in the process. Therefore, this chapter presents a method to obtain liveness-enforcing and optimal or near-optimal Petri net supervisors for RASs based on designing place invariants on critical activity places. We show that only the critical activity places need to be considered in designing place invariants to obtain such supervisors. The number of places considered in the supervisor synthesis is thus reduced, and the corresponding SSP becomes simpler.

However, its main limitation is that it still requires to compute the whole reachability graph as in Chen and Li (2011), Huang *et al.* (2015*b*, 2015*c*) and Uzam (2002) and it is thus not applicable to large-scale or unbounded Petri nets (Wang *et al.*, 2015; Lu *et al.*, 2019). Furthermore, the legal region of the system also needs to be convex and all designed place invariants are P-semiflows, i.e., place invariants with non-negative entries. So, efficient computation of legal markings and FBMs and expansion of the methods requires further research.

4.7 Bibliographical Notes

Reachability graph analysis is an important method used to synthesize Petri net supervisors for resource allocation systems. The theory of regions based on Petri net reachability graph analysis can be seen in Uzam (2002). Different approaches to design place invariants to permit legal markings and forbid illegal ones in the supervisor synthesis can be found in Chen and Li (2011), Cordone and Piroddi (2011), Huang *et al.* (2015*b*), Uzam and Zhou (2006, 2007) and Yamalidou *et al.* (1996). The original vector covering method is presented in (Chen & Li, 2011). The material reported in this chapter for reducing the number of places to be considered in the place invariant design mainly comes from our recent work (Huang *et al.*, 2019).

5

Redundant Constraint Elimination

A Minimal-number-of-Monitors Problem (MMP), which is presented in Section 3.3.4 to obtain optimal supervisors with the fewest monitors for resource allocation systems (RASs), needs extensive computation. Methods existing in the literature have mainly focused on the revision of the original formulation of MMP to reduce computational burden. This chapter presents another kind of methods that can be used to accelerate the MMP solution by eliminating its redundant reachability constraints. As a result, the problem scale for supervisor synthesis is drastically reduced. In this chapter, a sufficient and necessary condition for a reachability constraint to be redundant is established in the form of an integer linear program (ILP), based on the concept of feasible regions of supervisors. Then, two redundancy elimination methods are presented: an ILP one and a non-ILP one. Most of the redundant reachability constraints in MMP can be eliminated by these methods in a short time. Thus, the computational time to solve MMP is greatly reduced after the elimination, especially for large-scale models. Also, the obtained supervisors are still optimal and structurally minimal. Finally, numerical tests are conducted to show the efficiency and effectiveness of the presented methods.

5.1 Introduction

In Section 3.3.4, a reachability graph-based method is presented to obtain an optimal liveness-enforcing supervisor with the fewest monitors for DP-nets of resource allocation systems. The reachability graph of a DP-net is divided into a live zone and a deadlock zone. The markings in the live zone are legal ones that can reach the initial state of the underlying system while markings in the deadlock zone are deadlock markings, livelock markings, and bad markings that inevitably result in deadlocks and livelocks. First-met bad markings (FBMs) are markings

Supervisory Control and Scheduling of Resource Allocation Systems: Reachability Graph Perspective,
First Edition. Bo Huang and MengChu Zhou.

in the deadlock zone and they are immediately reachable by markings in the live zone. A vector covering method is adopted to reduce the number of markings in the sets of legal markings and FBMs to a minimal covering set $\mathcal{M}_L^*$ of legal markings and a minimal covered set $\mathcal{M}_F^*$ of FBMs, respectively. Then, an optimal supervisor with the fewest monitors can be obtained by solving a formulated MMP that forbids all markings in $\mathcal{M}_F^*$ and permits all markings in $\mathcal{M}_L^*$ when MMP has an optimal solution. This method can ensure that the obtained supervisor is optimal and structurally minimal in terms of the added monitors. However, it has a limitation: its computational burden can be extremely heavy, especially for large-scale models.

Existing acceleration techniques for it mainly focus on revising the formulation of MMP. For example, Chen *et al.* (2012) propose an iterative strategy to design an optimal supervisor based on place invariants. At each iteration, a Maximal-number-of-Forbidden FBMs Problem (MFFP) is solved to forbid as many FBMs as possible while keeping all legal markings of the model. It can reduce the computational time greatly, but it cannot guarantee the structural minimality of the obtained supervisor. To find an optimal and structurally minimal supervisor quickly, Chen and Li (2012) propose a Minimal-number-of-P-semiflows Problem (MPP) that has fewer constraints and variables than MMP. However, it has an initially undecidable parameter, i.e., the number of place invariants n_I, to be computed, and the efficiency of the method greatly depends on the initial selection of n_I. Moreover, if n_I is set to be less than the minimal number of monitors of the problem, MPP fails to obtain any solution.

In MMP, we see that an integer linear program (ILP) must be formulated and solved to obtain a supervisor. Yet, the methods that accelerate the ILP solution by eliminating its redundant constraints have seldom been investigated. Huang *et al.* (2015a) develop a method to eliminate redundant constraints of an ILP in the supervisor synthesis. However, it is conducted in the context of designing an optimal Petri net supervisor with self-loops. This chapter presents a method to simplify the MMP's formulation by eliminating redundant reachability constraints. In fact, most of the constraints in MMP are reachability ones and many of them are redundant in supervisor synthesis. If they were eliminated, the scale of the MMP would be greatly reduced, thereby speeding up a supervisor synthesis process. To define redundant reachability constraints, we first introduce the concept of a feasible region of supervisors. It is defined as the set of all feasible combinations of monitors for which all constraints in MMP are satisfied. Based on it, a reachability constraint in MMP is said to be redundant if it can be eliminated without changing the feasible region of supervisors. Since a redundant constraint is inactive for all feasible supervisors, its elimination does not change the optimal solution of MMP.

5.2 Minimal-Number-of-Monitors Problem

The MMP given in Section 3.3.4 deals with the structure optimization of optimal supervisors for DP-nets. In supervisor synthesis, a minimal covering set $\mathcal{M}_L^*$ of legal markings and a minimal covered set $\mathcal{M}_F^*$ of FBMs are calculated by a vector covering method, which is presented in Section 3.3.3. To compactly describe and conveniently analyze the MMP, this chapter uses the following notations.

X — An $(|\mathcal{M}_F^*| \cdot |\mathcal{M}_L^*|) \times |P_A|$ integer matrix of marking differences between $M_l \in \mathcal{M}_L^*$ and $M_j \in \mathcal{M}_F^*$, whose entry is $x_{l,j}(p_i) = M_l(p_i) - M_j(p_i)$ with $i \in \{1, 2, \dots, |P_A|\}$.

$x_{l,j}$ — A row of X (a $|P_A|$-dimensional integer vector) with $l \in \{1, 2, \dots, |\mathcal{M}_L^*|\}$ and $j \in \{1, 2, \dots, |\mathcal{M}_F^*|\}$.

$X^{<*,j>}$ — A matrix obtained by including the rows related to $M_j \in \mathcal{M}_F^*$ in X.

$\tilde{X}^{<l,j>}$ — A matrix obtained by eliminating $x_{l,j}$ from X with $l \in \{1, 2, \dots, |\mathcal{M}_L^*|\}$ and $j \in \{1, 2, \dots, |\mathcal{M}_F^*|\}$.

$\tilde{X}^{<l,*,j>}$ — A matrix obtained by eliminating $x_{l,j}$ and the rows related to the found redundant constraints from $X^{<*,j>}$.

$\tilde{X}_u^{<l,*,j>}$ — The uth row of $\tilde{X}^{<l,*,j>}$.

E — An $(|\mathcal{M}_F^*| \cdot (|\mathcal{M}_F^*| - 1)) \times |P_A|$ integer matrix of marking differences between $M_k \in \mathcal{M}_F^*$ and $M_j \in \mathcal{M}_F^*$ ($k \neq j$), whose entry is $e_{k,j}(p_i) = M_k(p_i) - M_j(p_i)$ with $i \in \{1, 2, \dots, |P_A|\}$.

$e_{k,j}$ — A row of E and a $|P_A|$-dimensional vector of marking differences between M_k and M_j in $\mathcal{M}_F^*$ such that $k, j \in \{1, 2, \dots, |\mathcal{M}_F^*|\}$ and $k \neq j$.

$f_{j,k}$ — A binary variable such that $f_{j,k} = 1$ if $M_k \in \mathcal{M}_F^*$ is forbidden by a place invariant designed to forbid $M_j \in \mathcal{M}_F^*$; otherwise $f_{j,k} = 0$.

z_j — A $|P_A|$-dimensional non-negative integer vector of the coefficients of the place invariant designed to forbid $M_j \in \mathcal{M}_F^*$.

q_j — A binary variable such that $q_j = 1$ if the place invariant designed to forbid $M_j \in \mathcal{M}_F^*$ is selected to compute a monitor; otherwise $q_j = 0$.

Γ — A positive integer that is chosen big enough.

$\gamma_{l,j}$ — A reachability constraint: $x_{l,j} \cdot z_j^T \leqslant -1$ with $l \in \{1, 2, \dots, |\mathcal{M}_L^*|\}$ and $j \in \{1, 2, \dots, |\mathcal{M}_F^*|\}$.

$S(X, E)$ — The system defined by the constraints in MMP.

$\Omega(X, E)$ — The feasible region of all possible supervisors in $S(X, E)$.

$$\min \sum_{j \in \{1, 2, \dots, |\mathcal{M}_F^*|\}} q_j$$

subject to

$$x_{l,j} \cdot z_j^T \leqslant -1, \ \forall l \in \{1, 2, \dots, |\mathcal{M}_L^*|\} \text{ and } \forall j \in \{1, 2, \dots, |\mathcal{M}_F^*|\}$$

$$e_{k,j} \cdot z_j^T \geq \Gamma \cdot (f_{j,k} - 1), \ \forall j, k \in \{1, 2, \ldots, |\mathcal{M}_F^*|\} \text{ and } j \neq k$$

$$f_{j,k} \leq q_j, \ \forall j, k \in \{1, 2, \ldots, |\mathcal{M}_F^*|\} \text{ and } j \neq k$$

$$q_j + \sum_{k \in \{1, 2, \ldots, |\mathcal{M}_F^*|\}, k \neq j} f_{k,j} \geq 1, \ \forall j \in \{1, 2, \ldots, |\mathcal{M}_F^*|\}. \tag{5.1}$$

At first sight, the above MMP formulation is different from the one in Section 3.3.4. However, they are in essence the same. In this formulation, reachability constraints are expressed by the inequalities containing $x_{l,j}$.

5.3 Elimination of Redundant Constraints

In this section, the definition of a redundant reachability constraint is given. If most reachability constraints in MMP are redundant and eliminated efficiently, the solution of MMP and the whole supervisor synthesis process may be considerably accelerated. Then, two elimination methods are presented and a procedure to combine the constraint elimination with the supervisor synthesis is given.

5.3.1 Redundant Reachability Constraints

The constraints in (5.1) determine the feasible region of optimal Petri net supervisors for the underlying system. For generalization, we remove the objective function from (5.1) and consider the following resultant system that is denoted by $S(X, E)$.

System $S(X, E)$:

$$x_{l,j} \cdot z_j^T \leq -1, \ \forall l \in \{1, 2, \ldots, |\mathcal{M}_L^*|\} \text{ and } \forall j \in \{1, 2, \ldots, |\mathcal{M}_F^*|\}$$

$$e_{k,j} \cdot z_j^T \geq \Gamma \cdot (f_{j,k} - 1), \ \forall j, k \in \{1, 2, \ldots, |\mathcal{M}_F^*|\} \text{ and } j \neq k$$

$$f_{j,k} \leq q_j, \ \forall j, k \in \{1, 2, \ldots, |\mathcal{M}_F^*|\} \text{ and } j \neq k$$

$$q_j + \sum_{k \in \{1, 2, \ldots, |\mathcal{M}_F^*|\}, k \neq j} f_{k,j} \geq 1, \ \forall j \in \{1, 2, \ldots, |\mathcal{M}_F^*|\}, \tag{5.2}$$

where X denotes an $(|\mathcal{M}_F^*| \cdot |\mathcal{M}_L^*|) \times |P_A|$ integer matrix whose element is $x_{l,j}(p_i)$ that represents the marking difference between $M_l \in \mathcal{M}_L^*$ and $M_j \in \mathcal{M}_F^*$ in terms of the ith activity place, and E denotes an $(|\mathcal{M}_F^*| \cdot (|\mathcal{M}_F^*| - 1)) \times |P_A|$ integer matrix whose entry $e_{k,j}(i)$ represents the marking difference between two different markings $M_k, M_j \in \mathcal{M}_F^*$ in terms of the ith activity place. For different DP-nets, X and E may be different. Therefore, such a system is denoted by $S(X, E)$. Before defining redundant reachability constraints, we use $\Omega(X, E)$, a feasible region of supervisors in $S(X, E)$, to represent all possible supervisors satisfying (5.2) and $\tilde{X}^{<m,n>}$ to denote the matrix obtained by deleting $x_{m,n}$ ($m \in \{1, 2, \ldots, |\mathcal{M}_L^*|\}$, $n \in \{1, 2, \ldots, |\mathcal{M}_F^*|\}$) from X.

Definition 5.1   *A reachability constraint*

$$x_{m,n} \cdot z_n^T \leqslant -1 \tag{5.3}$$

is redundant in $S(X, E)$ if

$$\Omega(X, E) = \Omega(\tilde{X}^{<m,n>}, E). \tag{5.4}$$

In other words, if the feasible region of a system remains unchanged after a reachability constraint is eliminated, then the constraint is said to be redundant. Note that the redundancy is not defined for any specific function or solution, but for all feasible optimal supervisors in $S(X, E)$. Condition (5.4) implies that the following equations are satisfied:

$$\Omega(X, E) \subseteq \Omega(\tilde{X}^{<m,n>}, E) \tag{5.5}$$

$$\Omega(X, E) \supseteq \Omega(\tilde{X}^{<m,n>}, E). \tag{5.6}$$

Condition (5.5) is satisfied since the removal of a reachability constraint only expands the feasible region. Thus, (5.6) is the center of our focus.

5.3.2   Linear Program Method

In this section, a sufficient and necessary condition for a reachability constraint to be redundant is given based on an ILP problem. Then, two simpler sufficient conditions are derived. Based on the simplest sufficient condition, a redundancy elimination algorithm for MMP is given.

Theorem 5.1   *Consider the following ILP problem*

$$\mathbb{P}_{m,n}^1 = \max x_{m,n} \cdot z_n^T$$

subject to

$$x_{l,j} \cdot z_j^T \leqslant -1, \ \forall l \in \{1, 2, \ldots, |\mathcal{M}_L^*|\}, \ \forall j \in \{1, 2, \ldots, |\mathcal{M}_F^*|\}$$
$$\text{and } (l, j) \neq (m, n)$$
$$e_{k,j} \cdot z_j^T \geqslant \Gamma \cdot (f_{j,k} - 1), \ \forall j, k \in \{1, 2, \ldots, |\mathcal{M}_F^*|\} \text{ and } j \neq k$$
$$f_{j,k} \leqslant q_j, \ \forall j, k \in \{1, 2, \ldots, |\mathcal{M}_F^*|\} \text{ and } j \neq k$$
$$q_j + \sum_{k \in \{1, 2, \ldots, |\mathcal{M}_F^*|\}, k \neq j} f_{k,j} \geqslant 1, \ \forall j \in \{1, 2, \ldots, |\mathcal{M}_F^*|\}. \tag{5.7}$$

The constraint expressed in (5.3) is redundant in $S(X, E)$ iff $\mathbb{P}_{m,n}^1 \leqslant -1$.

Proof: First, we prove that if $\mathbb{P}^1_{m,n} \leqslant -1$, (5.3) is redundant in $S(X, E)$. In (5.7), all feasible solutions of this system constitute $\Omega(\tilde{X}^{<m,n>}, E)$. If $\mathbb{P}^1_{m,n} \leqslant -1$, then for any solution within $\Omega(\tilde{X}^{<m,n>}, E)$, we have:

$$x_{m,n} \cdot z_n^T \leqslant \mathbb{P}^1_{m,n} \leqslant -1 \tag{5.8}$$

which is the same as the eliminated reachability constraint, i.e., $x_{m,n} \cdot z_n^T \leqslant -1$. It means that a feasible solution to system $S(\tilde{X}^{<m,n>}, E)$ is also feasible to system $S(X, E)$. Therefore, (5.6) holds. According to Definition 5.1, (5.3) is redundant in $S(X, E)$.

Then, we prove that if (5.3) is a redundant constraint, $\mathbb{P}^1_{m,n} \leqslant -1$ holds. If the constraint in (5.3) is redundant, (5.6) should be true. So, any solution in $\Omega(\tilde{X}^{<m,n>}, E)$ is also a solution to $S(X, E)$. Since (5.3) is satisfied by all solutions to $S(X, E)$, the maximal value of the objective function, which is the same as the left side of (5.3), should be less than -1. Therefore, $\mathbb{P}^1_{m,n} \leqslant -1$ holds. ∎

Theorem 5.1 gives a sufficient and necessary condition for a reachability constraint to be redundant in $S(X, E)$. However, the ILP problem $\mathbb{P}^1_{m,n}$ given in Theorem 5.1 has $|\mathcal{M}^*_F| \cdot (|\mathcal{M}^*_L| + 2|\mathcal{M}^*_F| - 1) - 1$ constraints and $|\mathcal{M}^*_F| \cdot (|P_A| + |\mathcal{M}^*_F|)$ variables. It has little practical use due to the enormous amounts of time required by a series of ILPs that eliminate redundancy in MMP.

Theorem 5.2 *Consider the following ILP problem*

$$\mathbb{P}^2_{m,n} = \max x_{m,n} \cdot z_n^T$$

subject to

$$x_{l,j} \cdot z_j^T \leqslant -1, \ \forall l \in \{1, 2, \ldots, |\mathcal{M}^*_L|\}, \ \forall j \in \{1, 2, \ldots, |\mathcal{M}^*_F|\}$$

$$\text{and } (l, j) \neq (m, n). \tag{5.9}$$

A reachability constraint expressed in (5.3) is redundant in $S(X, E)$ if $\mathbb{P}^2_{m,n} \leqslant -1$.

Proof: In (5.9), all constraints except the reachability constraints with $(l, j) \neq (m, n)$ are taken away from (5.7). So, the feasible region of problem (5.9) is a superset of that of problem (5.7), which implies that $\mathbb{P}^2_{m,n} \geqslant \mathbb{P}^1_{m,n}$. Therefore, if $\mathbb{P}^2_{m,n} \leqslant -1$, then $\mathbb{P}^1_{m,n} \leqslant -1$ and (5.3) is thus redundant in $S(X, E)$ according to Theorem 5.1. ∎

Theorem 5.2 provides a simpler sufficient condition for a reachability constraint to be redundant. Now, we give the Petri net interpretation for it. In MMP, only activity places are considered when constructing a place invariant to prevent an FBM from being reached (Uzam & Zhou, 2006). To forbid an FBM, the weighted sum of tokens in activity places of the place invariant should be one less than that of the FBM and greater than or equal to that of any legal marking. The objective function in (5.9) can be interpreted as the maximal possible difference between the weighted sum of tokens in the activity places of a legal marking and that of an

FBM for all reachability constraints except (5.3). If the maximal possible difference is still less than or equal to the limit, i.e., -1, the reachability constraint expressed in (5.3) is undoubtedly redundant.

The ILP problem $\mathbb{P}^2_{m,n}$ given in Theorem 5.2 has $|\mathcal{M}^*_F| \cdot |\mathcal{M}^*_L| - 1$ constraints and $|\mathcal{M}^*_F| \cdot |P_A|$ variables, which are fewer than those of $\mathbb{P}^1_{m,n}$. In the following, we present an MMP redundancy elimination algorithm based on an ILP problem $\mathbb{P}^3_{m,n}$, which is revised from $\mathbb{P}^2_{m,n}$ and has fewer constraints and variables than $\mathbb{P}^2_{m,n}$ to judge the redundancy of each reachability constraint. Let $X^{<*,n>}$ be a matrix including all rows related to $M_n \in \mathcal{M}^*_F$ in X and $\tilde{X}^{<m,*,n>}$ be a matrix obtained by removing $x_{m,n}$ and the rows related to the found redundant reachability constraints from $X^{<*,n>}$.

$$\mathbb{P}^3_{m,n} = \max x_{m,n} \cdot z^T_n$$

subject to

$$\tilde{X}^{<m,*,n>} \cdot z^T_n \leqslant -1.$$

Algorithm 5.1

```
Redundant constraint elimination by using an ILP method.
Input: The MMP of a DP-net.
Output: A reduced MMP.
Redundancy_elimination_ILP {
        R := ∅; /*R denotes the set of redundant constraints found in
the MMP.*/
        for each constraint γ_{m,n} : x_{m,n} · z^T_n ⩽ -1 in the MMP
        {
                Q := X^{<*,n>}; /*Q is used to obtain X̃^{<m,*,n>}.*/
                Q := Q̃^{<m,n>}; /*Delete x_{m,n} from Q.*/
                for each γ_{l,n} ∈ R
                        Q := Q̃^{<l,n>}; /*Delete x_{l,n} from Q.*/
                X̃^{<m,*,n>} := Q;
                Solve the following ILP problem
                        P³_{m,n} = max    x_{m,n}  ·  z^T_n
                        subject to
                                X̃^{<m,*,n>} ·  z^T_n ⩽ -1
                if P³_{m,n} ⩽  -1
                        R := R ∪ {γ_{m,n}};
        }
        MMP := MMP \ R;
        Output MMP.
}
```

Algorithm 5.1 identifies and eliminates redundant reachability constraints from MMP by using $\mathbb{P}^3_{m,n}$. For each reachability constraint $\gamma_{m,n}$ in MMP, it first selects the reachability constraints related to $M_n \in \mathcal{M}^*_F$ as the candidate constraints of $\mathbb{P}^3_{m,n}$. Next, $\gamma_{m,n}$ and all M_n-related redundant constraints are removed. Then, with the remaining constraints, $\mathbb{P}^3_{m,n}$ is solved. If the result is less than or equal to -1, $\gamma_{m,n}$ is redundant. Finally, all found redundant constraints are removed to obtain the

Table 5.1 Numbers of constraints and variables in different ILPs.

ILP	Number of constraints	Number of variables
$\mathbb{P}^1_{m,n}$	$\|\mathcal{M}^*_F\| \cdot (\|\mathcal{M}^*_L\| + 2\|\mathcal{M}^*_F\| - 1) - 1$	$\|\mathcal{M}^*_F\| \cdot (\|P_A\| + \|\mathcal{M}^*_F\|)$
$\mathbb{P}^2_{m,n}$	$\|\mathcal{M}^*_F\| \cdot \|\mathcal{M}^*_L\| - 1$	$\|\mathcal{M}^*_F\| \cdot \|P_A\|$
$\mathbb{P}^3_{m,n}$	$\leqslant \|\mathcal{M}^*_L\| - 1$	$\|P_A\|$

Source: Adapted from Huang *et al.* (2015*b*). Reproduced with permission of Elsevier.

reduced MMP. The advantages of the method are that the numbers of constraints and variables needed to judge the redundancy of a constraint are at most $\|\mathcal{M}^*_L\| - 1$ (since some redundant rows may be removed from $\tilde{X}^{<m,*,n>}$) and $\|P_A\|$, respectively, which can be seen in Table 5.1, and all the constraints eliminated by Algorithm 5.1 are redundant.

Theorem 5.3 *All constraints eliminated by Algorithm 5.1 are redundant in* $S(X, E)$.

Proof: In Algorithm 5.1, the constraints in $\mathcal{R}$ are removed. Now, we prove that $\forall \gamma_{m,n} \in \mathcal{R}$, it is redundant in $S(X, E)$. $\forall \gamma_{m,n} \in \mathcal{R}$, we have that $\mathbb{P}^3_{m,n} \leqslant -1$. In the problem $\mathbb{P}^3_{m,n}$, the reachability constraints not related to $M_n \in \mathcal{M}^*_F$ and the redundant constraints already found are removed when compared with $\mathbb{P}^2_{m,n}$. Therefore, the feasible region of $\mathbb{P}^3_{m,n}$ is a superset of that of $\mathbb{P}^2_{m,n}$. Thus, we have that $\mathbb{P}^2_{m,n} \leqslant \mathbb{P}^3_{m,n}$, which implies that $\mathbb{P}^2_{m,n} \leqslant -1$. Based on Theorem 5.2, $\gamma_{m,n}$ is redundant in $S(X, E)$. ∎

5.3.3 Non-Linear Program Method

In Algorithm 5.1, $\mathbb{P}^3_{m,n}$ is still an ILP problem, but with fewer constraints and variables. It would be desirable to develop a method that identifies redundant constraints without solving any optimization problem. This subsection presents such a method.

Theorem 5.4 *A reachability constraint* $x_{m,n} \cdot z^T_n \leqslant -1$ *is redundant in the system* $S(X, E)$ *if there exists at least one other constraint*

$$x_{l_1,n} \cdot z^T_n \leqslant -1,$$

$$x_{l_2,n} \cdot z^T_n \leqslant -1,$$

$$\vdots$$

$$x_{l_w,n} \cdot z^T_n \leqslant -1 \tag{5.10}$$

that satisfies $w \geqslant 1$ *and* $\forall i \in \{1, 2, \ldots, \|P_A\|\}$, $x_{m,n}(p_i) \leqslant \sum^w_{k=1} x_{l_k,n}(p_i)$.

Proof: *By adding up both sides of the constraints in (5.10), we have* $\sum_{k=1}^{w} x_{l_k,n} \cdot z_n^T \leqslant -w$ *where* w *is the number of these constraints. Since* $\forall i \in \{1, 2, \ldots, |P_A|\}$, $x_{m,n}(p_i) \leqslant \sum_{k=1}^{w} x_{l_k,n}(p_i)$ *and all elements in* z_n^T *are non-negative, we have that* $x_{m,n} \cdot z_n^T \leqslant \sum_{k=1}^{w} x_{l_k,n} \cdot z_n^T$. *Therefore, we can derive* $x_{m,n} \cdot z_n^T \leqslant -w \leqslant -1$, *which means that the reachability constraint* $x_{m,n} \cdot z_n^T \leqslant -1$ *is satisfied in* $S(\tilde{X}^{<m,n>}, E)$. *It also indicates that any feasible solution in* $\Omega(\tilde{X}^{<m,n>}, E)$ *is feasible to system* $S(X, E)$. *Therefore, (5.6) is satisfied. According to Definition 5.1, (5.3) is redundant in* $S(X, E)$. ∎

Based on Theorem 5.4, we present a redundancy elimination method for the MMP in Algorithm 5.1 in which $\tilde{X}_u^{<m,*,n>}$ denotes the uth row of $\tilde{X}^{<m,*,n>}$ and $|\mathcal{M}_L^*|$ is the number of markings in $\mathcal{M}_L^*$.

Algorithm 5.2

```
Redundant constraint elimination by using a non-ILP method.
Input: The MMP of a DP-net.
Output: A reduced MMP.
Redundancy_elimination_nonILP {
1        R := ∅; /*R denotes the set of redundant constraints found in the
MMP.*/
2        for each constraint γ_{m,n} : x_{m,n} · z_n^T ≤ -1
         {
3            Q := X^{<*,n>}; /*Q is used to obtain X̃^{<m,*,n>}.*/
4            Q := Q̃^{<m,n>}; /*Delete x_{m,n} from Q.*/
5            for each γ_{l,n} ∈ R
6                Q := Q̃^{<l,n>}; /*Delete x_{l,n} from Q.*/
7            X̃^{<m,*,n>} := Q;
8            for u := 1 to |M_L^*| - |R| - 1
9                for v := u+1 to |M_L^*| - |R| - 1
10                   if ∀ i ∈ {1,2,···,|P_A|}, x_{m,n}(p_i) ≤ X̃_u^{<m,*,n>}(p_i) + X̃_v^{<m,*,n>}(p_i)
                     {
11                       R := R ∪ {γ_{m,n}};
12                       Go to Step 2 for the next constraint;
                     }
         }
13       MMP := MMP \ R;
14       Output MMP.
}
```

Algorithm 5.2 can identify and eliminate redundant reachability constraints from MMP by using the non-ILP method. For each reachability constraint $\gamma_{m,n}$, it only uses the constraints that have the same variables as $\gamma_{m,n}$ to judge the redundancy of $\gamma_{m,n}$. The procedures are as follows. First, the reachability constraints related to $M_n \in \mathcal{M}_F^*$ are selected as candidate constraints. Next, $\gamma_{m,n}$ and all redundant constraints found so far are removed from the candidates. Then, the remaining constraints are searched to find whether there exist two constraints that imply $\gamma_{m,n}$ or not. If yes, $\gamma_{m,n}$ is redundant. At the end of the algorithm, all found redundant constraints are deleted from MMP. The advantages of this method are that it does not need to solve any optimization problem and all the constraints eliminated are redundant.

Theorem 5.5 *All constraints eliminated by Algorithm 5.2 are redundant in system $S(X, E)$.*

Proof: In Algorithm 5.2, the constraints in $\mathcal{R}$ are the ones to be eliminated. $\forall \gamma_{m,n} \in \mathcal{R}$, there exist two constraints $\tilde{X}_u^{<m,*,n>} \cdot z_n^T \leqslant -1$ and $\tilde{X}_v^{<m,*,n>} \cdot z_n^T \leqslant -1$ such that $\forall i \in \{1, 2, \dots, |P_A|\}$, $x_{m,n}(p_i) \leqslant \tilde{X}_u^{<m,*,n>}(p_i) + \tilde{X}_v^{<m,*,n>}(p_i)$. Since $\tilde{X}^{<m,*,n>} \subseteq \tilde{X}^{<m,n>}$, there exist two constraints $x_{l_1,n} \cdot z_n^T \leqslant -1$ and $x_{l_2,n} \cdot z_n^T \leqslant -1$ with $x_{l_1,n} = \tilde{X}_u^{<m,*,n>} \in \tilde{X}^{<m,n>}$ and $x_{l_2,n} = \tilde{X}_v^{<m,*,n>} \in \tilde{X}^{<m,n>}$ such that $\forall i \in \{1, 2, \dots, |P_A|\}$, $x_{m,n}(p_i) \leqslant \sum_{k=1}^{2} x_{l_k,n}(p_i)$. According to Theorem 5.4, $\gamma_{m,n}$ is redundant in $S(X, E)$. ∎

5.3.4 Supervisor Synthesis with Redundancy Elimination

This section presents a deadlock prevention method by using MMP with redundancy elimination to obtain an optimal supervisor with the fewest monitors.

Algorithm 5.3
```
Deadlock prevention method.
Input: Petri net model (N, M₀) of a DP-net.
Output: A controlled Petri net system.
Deadlock_prevention {
        Generate the reachability graph G(N, M₀) of the net and
compute M_F* and M_L*;
        A := ∅; /*A denotes the set of monitors to be added.*/
        Formulate the MMP for the net model;
        Compute a reduced MMP by using Algorithm 5.1 or
Algorithm 5.2;
        Solve the reduced MMP. If it has no solution, exit;
        for each j satisfying qⱼ = 1
        {
             Use zⱼ in the solution as the coefficients of a place
invariant and design a monitor aⱼ by using the method presented in
Section 3.3.2;
             A := A ∪ {aⱼ};
        }
        Add all monitors in A to the original net and output the
controlled net;
}
```

Algorithm 5.3 first generates the markings in $\mathcal{M}_F^*$ and $\mathcal{M}_L^*$ of $G(N, M_0)$ and formulates MMP. Next, an ILP or non-ILP method is used to identify and eliminate redundant reachability constraints in MMP. Note that the elimination of redundant constraints does not change the feasible region of the problem. Then, monitors that contain control places and arcs in a supervisor can be obtained by solving the reduced MMP. The obtained supervisor forbids all FBMs and permits all legal markings with the fewest monitors. In addition, we have the following conclusion.

Theorem 5.6 *Algorithm 5.3 obtains an optimal supervisor with the fewest monitors for a DP-net if the reduced MMP has an optimal solution.*

Proof: In the reduced MMP, only the redundant constraints found by Algorithm 5.1 or Algorithm 5.2 are removed from the original MMP. Therefore, the original MMP and the reduced one are in essence the same since the elimination of redundant constraints does not change the feasible region of the MMP. If the reduced MMP has an optimal solution, all FBMs are forbidden by the solution. This is because $\forall M_j \in \mathcal{M}_F^*$ is forbidden by at least one designed place invariant according to the constraints $q_j + \sum_{k \in \{1,2,\ldots,|\mathcal{M}_F^*|\}, k \neq j} f_{k,j} \geqslant 1$, $\forall j \in \{1, 2, \ldots, |\mathcal{M}_F^*|\}$, indicating that the obtained controlled net cannot enter the deadlock zone. On the other hand, all legal markings in $G(N, M_0)$ are permitted in the controlled net via the reachability constraints $x_{l,j} \cdot z_j^T \leqslant -1$, $\forall j \in \{1, 2, \ldots, |\mathcal{M}_F^*|\}$, $\forall l \in \{1, 2, \ldots, |\mathcal{M}_L^*|\}$. Thus, the resultant supervisor is optimal. According to the objective of MMP, the number of monitors in the supervisor is minimized. Therefore, the obtained supervisor is optimal and has the fewest monitors. ∎

Note that since the coefficients $z_j (j \in \{1, 2, \ldots, |\mathcal{M}_F^*|\})$ of the designed place invariants are non-negative, these place invariants are P-semiflows. In the next section, some experiments are tested to show that the proposed methods can eliminate many redundant constraints of MMP in a short time, thus accelerating the solution of MMP, especially for large-scale problems.

5.4 Illustrative Examples

In this section, some widely used resource allocation systems are tested in a computer with Intel i3 Core 2.93 GHz CPU and 4GB memory to show the effectiveness of the presented methods. Three cases are performed for each system. In Case 1, the standard MMP in Section 3.3.4 is conducted. Case 2 includes the fastest ILP elimination method presented in Algorithm 5.1. In Case 3, the non-ILP elimination method presented in Algorithm 5.2 is applied.

First, consider a flexible manufacturing system (Li *et al.*, 2008a; Uzam, 2002) whose DP-net is shown in Figure 5.1. It has 19 places and 14 transitions. The set of activity places is $\{p_2 - p_7, p_9 - p_{13}\}$. There are 282 markings in its reachability graph $G(N, M_0)$, 205 of which are legal markings and 54 of which are FBMs. By using the vector covering method, $\mathcal{M}_L^*$ and $\mathcal{M}_F^*$ have 26 and eight markings, respectively. For this net, the formulated MMP has 328 constraints, 208 of which

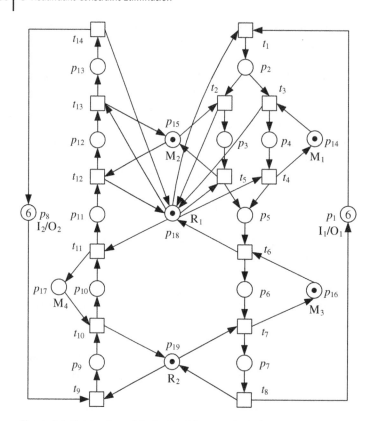

Figure 5.1 Petri net model of a flexible manufacturing system.

Table 5.2 Results for the model in Figure 5.1.

	Case 1	Case 2	Case 3
N_E	—	60	60
T_E	—	0.49 s	0.04 s
T_{MMP}	⩽1 s	⩽1 s	⩽1s

Source: Adapted from Huang *et al.* (2015b).
Reproduced with permission of Elsevier.

are reachability constraints. The numbers of eliminated constraints and the solution time of each method are shown in Table 5.2 in which N_E represents the number of eliminated constraints, T_E indicates the elimination time, and T_{MMP} denotes the time required for solving the resultant MMP. Although 60 redundant constraints are eliminated by the ILP and non-ILP method, no reduction of the MMP solution time is observed. This can be explained by the relatively small scale of the problem.

Table 5.4 shows the results of Algorithm 5.3 and the methods available in the literature regarding the number of monitors in the supervisor and the number of reachable markings in the controlled net. All the methods can obtain an optimal supervisor with which 205 legal states are all reachable in the controlled net. In addition, the approach presented in this chapter and those in Chen and Li (2011, 2012) can lead to an optimal supervisor with only two monitors, i.e., the fewest monitors in the supervisor.

Next, a more complex system (Chen et al., 2011, 2012; Ezpeleta et al., 1995; Huang et al., 2001; Piroddi et al., 2008) is tested. Its DP-net model is shown in Figure 5.2 that has 26 places and 20 transitions. The set of activity places is $\{p_2 - p_4, p_6 - p_{13}, p_{15} - p_{19}\}$. There are 26 750 reachable markings in its reachability graph $G(N, M_0)$, 21 581 of which are legal markings and 4 211 of which are FBMs. By using the vector covering method, $\mathcal{M}_L^*$ and $\mathcal{M}_F^*$ have 393 and 34 markings, respectively. For this model, the formulated MMP has 15 640 constraints, 13 362 of which are reachability ones. Table 5.3 shows that more than 57% of the constraints are eliminated by the ILP elimination method and more than 54% by the non-ILP one, both taking about two minutes only. In addition, a striking result is that after the redundancy eliminations, the MMP solution time is dramatically reduced from 30 hours and 29 minutes to 13 hours and 53 minutes and 10 hours and 41 minutes, respectively.

Table 5.5 shows the state-of-the-art results for the net shown in Figure 5.2 regarding the number of monitors in the supervisor and the number of reachable markings in the controlled net. We can see that the supervisors obtained by the approach presented in this chapter and the method in Chen and Li (2011) are optimal and have the fewest monitors. Moreover, the structure of the supervisor obtained by the approach of this chapter is guaranteed to be minimal in terms of the added monitors.

To further demonstrate the presented approach, the manufacturing example shown in Figure 5.2 is parameterized to generate different problem instances by varying the number of parts of each part type (tokens in p_5, p_{14}, and p_1) and the number of available units of each resource type (tokens in p_{23}, p_{24}, p_{25}, and p_{26}). Table 5.6 summarizes the results of the instances with a quintuplet $(M_0(p_5), M_0(p_{14}), M_0(p_1), M_0(p_{23}) = M_0(p_{25}), M_0(p_{24}) = M_0(p_{26}))$ in terms of the number of markings in $\mathcal{M}_F^*$ ($|\mathcal{M}_F^*|$), the number of markings in $\mathcal{M}_L^*$ ($|\mathcal{M}_L^*|$), the number of legal markings in the plant net (N_L), the number of monitors in the supervisor (N_M), the number of reachable marking in the controlled net (N_R), the number of constraints in the MMP (N_C), the number of reachability constraints (N_C'), the number of eliminated constraints (N_E), the elimination time ($\mathcal{T}_E$), and the solution time of the MMP ($\mathcal{T}_{MMP}$). The following phenomena are observed.

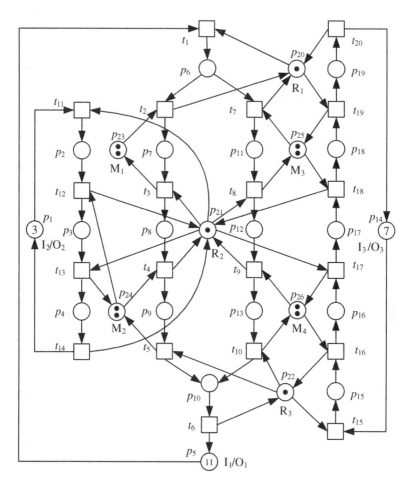

Figure 5.2 Petri net model of another manufacturing system.

Table 5.3 Results for the model in Figure 5.2.

	Case 1	Case 2	Case 3
N_E	—	8 939	8 461
$\mathcal{T}_E$	—	2 min 7 s	2 min 8 s
$\mathcal{T}_{MMP}$	30 h 29 min 7 s	13 h 53 min 58 s	10 h 41 min 42 s

Source: Adapted from Huang *et al.* (2015*b*). Reproduced with permission of Elsevier.

Table 5.4 Performance comparisons for the model in Figure 5.1.

	Uzam (2002)	Li *et al.* (2008a)	Piroddi *et al.* (2008)	Chen *et al.* (2011)	Chen and Li (2012)	Chen and Li (2011)	Algorithm 5.3
Monitors	6	9	5	8	2	2	2
Markings	205	205	205	205	205	205	205

Source: Adapted from Huang *et al.* (2015b). Reproduced with permission of Elsevier.

Table 5.5 Performance comparisons for the model in Figure 5.2.

	Ezpeleta *et al.* (1995)	Huang *et al.* (2001)	Uzam and Zhou (2006)	Chen *et al.* (2011)	Piroddi *et al.* (2008)	Chen and Li (2011)	Algorithm 5.3
Monitors	18	16	19	17	13	5	5
Markings	6 287	12 656	21 562	21 581	21 581	21 581	21 581

Source: Adapted from Huang *et al.* (2015b). Reproduced with permission of Elsevier.

1) The ILP elimination method can eliminate slightly more constraints than the non-ILP one, but the solution time of the reduced MMP is not always a monotonically decreasing function of the number of eliminated constraints in the problem. This reason is that although the number of constraints has a significant impact on solving linear program problems, it is not the only factor in general. For instance, the primal-dual potential-reduction algorithm (Luenberger & Ye, 2008) requires $O(mn^2 \log(n/\varepsilon))$ arithmetic calculations in its linear program model, in which m and n are the numbers of variables and constraints, and ε is the maximal acceptable duality gap. Therefore, it is reasonable for a reduced MMP with slightly more constraints to be solved faster than the one with fewer constraints.

2) The supervisor obtained by the approach presented in this chapter is optimal since the number of reachable markings in the controlled net is the same as the number of the legal markings in the plant net. In addition, the structure of the obtained supervisors can be guaranteed to be minimal in terms of added monitors.

3) The results suggest that the approach presented in this chapter are more beneficial to large-scale systems. In Table 5.6, we can see that as the model size increases, more redundant constraints are eliminated and more solution time is saved. Meanwhile, the redundancy elimination time remains short.

Table 5.6 Solutions of parameterized problem instances.

	$|\mathcal{M}_F^*|$	$|\mathcal{M}_L^*|$	N_L	N_M	N_R	N_C	N_C'		N_E	T_E	T_{MMP}
(4 3 2 1 1)	13	129	883	4	883	2002	1677	Case 1	/	/	1 min 41 s
								Case 2	534	6.12 s	1 min 25 s
								Case 3	531	3.24 s	1 min 35 s
(6 4 2 1 1)	13	71	990	4	990	1248	923	Case 1	/	/	1 min 14 s
								Case 2	370	2.67 s	0 min 54 s
								Case 3	364	0.59 s	1 min 11 s
(6 6 3 2 1)	17	199	3789	5	3789	3944	3383	Case 1	/	/	47 min 22 s
								Case 2	1654	17.46 s	29 min 23 s
								Case 3	1633	11.77 s	23 min 38 s
(8 6 3 2 1)	17	126	3887	5	3887	2703	2142	Case 1	/	/	28 min 22 s
								Case 2	1157	8.50 s	16 min 34 s
								Case 3	1130	3.18 s	24 min 47 s
(10 6 3 2 2)	34	403	21 561	5	21 561	15 980	13 702	Case 1	/	/	30 h 59 min 38 s
								Case 2	9176	2 min 17 s	15 h 39 min 38 s
								Case 3	8671	2 min 19 s	18 h 51 min 10 s
(12 7 4 2 2)	34	393	21 581	5	21 581	15 640	13 362	Case 1	/	/	30 h 29 min 07 s
								Case 2	8939	2 min 07 s	13 h 53 min 58 s
								Case 3	8461	2 min 08 s	10 h 41 min 42 s

Source: Adapted from Huang *et al.* (2015*b*). Reproduced with permission of Elsevier.

5.5   Concluding Remarks

In MMP presented in Section 3.3.4, many reachability constraints are redundant. Thus, its solution and the related supervisor synthesis process can be accelerated if the redundant constraints can be efficiently eliminated. To address it, this chapter presents two techniques: an ILP-based method and a non-ILP method. Most of the redundant reachability constraints in MMP can be identified and removed by both methods in a short time. For most instances, the ILP methods can eliminate slightly more constraints than the non-ILP one, but the latter does not need to solve any optimization problem. The solving process of MMP is thus accelerated owing to the efficient elimination of redundant constraints, especially for large-scale systems. In addition, the obtained supervisors are maximally permissive in behavior and structurally minimal in structure.

5.6   Bibliographical Notes

The material of this chapter mainly comes from Huang *et al.* (2015*b*). For the original ILP used to synthesize the optimal supervisor, please refer to Chen and Li (2011). Compared with other acceleration methods for the solving of an ILP, such as the iterative method in Chen *et al.* (2012) and the approach with an alternative ILP in Chen and Li (2012), this chapter investigates a method that accelerates the solving process of ILP by efficiently eliminating its redundant constraints, while not affecting the quality of the obtained supervisor.

6

Fast Iterative Supervisor Design

This chapter investigates several acceleration techniques that can be used to synthesize optimal supervisors with a compressed structure for DP-nets of resource allocation systems (RASs). An optimal supervisor can forbid all first-met bad markings and permit all legal markings by some designed place invariants. This chapter presents an iterative method that formulates a lexicographic multiobjective integer linear program (ILP) at each iteration to synthesize an optimal supervisor with a simple structure in terms of both control places and added arcs. Instead of a single large IPL, several much smaller IPLs are formulated in the method, resulting in a much faster synthesis process. To further reduce the ILP solving time, a redundancy identification method is also adopted in the iterative method.

6.1 Introduction

To generate an optimal and structurally minimal supervisor for the DP-net of an RAS, Chapter 3 presents a method to design place invariants that permit all legal markings and forbid all first-met bad markings (FBMs). In order to do so, an integer linear program (ILP) whose objective is to minimize the number of control places in the supervisor is formulated and solved. The method can guarantee that the controlled net is optimal and the added supervisor has the fewest control places. However, in the real world, the implementation cost of a supervisor is evaluated in two parts: the implementation cost of control places and added arcs. A control place represents a processing unit such as a programmable logic controller, a microcontroller, or a computer. An added arc in a supervisor represents a sensor that collects data from an operation or an actuator that transmits signals to an operation. The approach in Chen *et al.* (2014a) considers the implementation cost of both control places and added arcs through an ILP. Huang *et al.* (2015c) also develop a method to design an optimal supervisor with the fewest added arcs while ensuring that the number of control places is

Supervisory Control and Scheduling of Resource Allocation Systems: Reachability Graph Perspective,
First Edition. Bo Huang and MengChu Zhou.
© 2020 The Institute of Electrical and Electronics Engineers, Inc. Published 2020 by John Wiley & Sons, Inc.

minimal by solving a lexicographic ILP. The advantage of these methods is that the obtained supervisors are optimal and structurally simple or minimal with regards to the numbers of added places and arcs. But the methods construct place invariants via a single ILP that is often too large to be solved for sizable models.

This chapter presents an iterative multiobjective linear program approach to quickly synthesize optimal Petri net supervisors for DP-nets of RASs. In addition, the method in Chapter 5 can efficiently eliminate redundant reachability constraints in an ILP. However, it is conducted in the context of optimal supervisor synthesis via one ILP. In this chapter, an iterative lexicographic multiobjective integer linear program (LMILP) with redundancy reduction is developed. The method has the following features: i) Instead of a single ILP, several much smaller LMILPs are formulated to design supervisors for DP-nets via an iterative approach. ii) A sufficient and necessary condition for a reachability constraint in an ILP to be redundant is established and an efficient redundancy identification method is given for the iterative method. Such redundancy elimination can further reduce the solution time of an LMILP. iii) An optimal supervisor with a compressed structure in terms of both control places and connected arcs can be obtained by the presented method.

The method can be applied to the DP-nets that comply with Definition 3.1, such as S^3PR, ES^3PR, LS^3PR, ELS^3PR, GLS^3PR, S^2LSPR, S^*PR, S^4PR, S^3PGR^2, and S^3PMR whose definitions can be seen in Section 2.2.5. But its application scope is smaller than that of the theory of regions (Ghaffari *et al.*, 2003) since the method is limited to Petri nets in Definition 3.1, which model sequential or non-sequential resource allocation systems.

6.2 Optimal Supervisor of a DP-net

According to Definition 3.1, places in a DP-net of an RAS can be classified into idle places P^0, resource places P_R, and activity places P_A. Tokens in an idle place denote the maximal number of concurrent operations that can be performed in a processing subnet. Resource places represent resources and their initial tokens stand for the number of available resource units at the very beginning. An activity (operation) place represents an operation to be processed in a job and it has no token in the initial marking M_0. For example, we consider an automated manufacturing system that has three kinds of shared resources R_1–R_3, two loading buffers I_1–I_2, and two unloading buffers O_1–O_2. A resource only processes one part at a time. Two types of parts, i.e., P_1 and P_2, are processed in the system. The production sequences are:

$$P_1: I_1 \rightarrow R_1 \rightarrow R_2 \rightarrow R_3 \rightarrow O_1$$
$$P_2: I_2 \rightarrow R_3 \rightarrow R_2 \rightarrow R_1 \rightarrow O_2$$

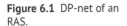

Figure 6.1 DP-net of an RAS.

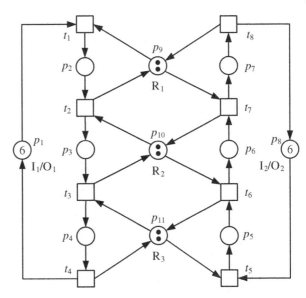

Figure 6.1 shows its DP-net that has 11 places and 8 transitions. The places are divided into $P^0 = \{p_1, p_8\}$, $P_R = \{p_9 - p_{11}\}$, and $P_A = \{p_2 - p_7\}$. Its reachability graph is given in Figure 6.2 where a compact multiset formalism $\sum_i M(p_i) \cdot p_i$ is used to denote each marking for conciseness. In the reachability graph, $M_{192} - M_{202}$ are FBMs and the rest are legal markings. By using the vector covering method in Section 3.3.3, the sets of legal markings and FBMs can be reduced into two smaller sets, i.e., a minimal covering set of legal markings $\mathcal{M}_L^*$ and a minimal covered set of FBMs $\mathcal{M}_F^*$. For this net, $\mathcal{M}_L^*$ only has 19 markings and $\mathcal{M}_F^* = \{2p_2 + p_3 + 2p_5 + p_6, 2p_3 + 2p_5, 2p_2 + 2p_6\}$. If all markings in $\mathcal{M}_L^*$ are permitted and all markings in $\mathcal{M}_F^*$ are forbidden by an added supervisor, the supervisor is said to be optimal and the resultant controlled net can run in the live zone and never enters the deadlock zone of the reachability graph. Such a supervisor can be computed by a place invariant-based method presented in Chapter 3.

6.3 Fast Synthesis of Optimal and Simple Supervisors

This section presents an efficient iterative deadlock prevention method to obtain optimal and structure-simple supervisors in terms of both the control places and added arcs. First, an LMILP is formulated at each iteration to obtain an optimal monitor with the objectives to forbid as many FBMs as possible in $\mathcal{M}_F^*$, generate the fewest arcs, and simplify the coefficient of the designed place invariant.

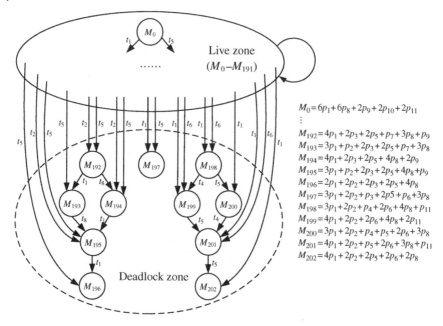

$M_0 = 6p_1 + 6p_8 + 2p_9 + 2p_{10} + 2p_{11}$

$\vdots$

$M_{192} = 4p_1 + 2p_3 + 2p_5 + p_7 + 3p_8 + p_9$
$M_{193} = 3p_1 + p_2 + 2p_3 + 2p_5 + p_7 + 3p_8$
$M_{194} = 4p_1 + 2p_3 + 2p_5 + 4p_8 + 2p_9$
$M_{195} = 3p_1 + p_2 + 2p_3 + 2p_5 + 4p_8 + p_9$
$M_{196} = 2p_1 + 2p_2 + 2p_3 + 2p_5 + 4p_8$
$M_{197} = 3p_1 + 2p_2 + p_3 + 2p5 + p_6 + 3p_8$
$M_{198} = 3p_1 + 2p_2 + p_4 + 2p_6 + 4p_8 + p_{11}$
$M_{199} = 4p_1 + 2p_2 + 2p_6 + 4p_8 + 2p_{11}$
$M_{200} = 3p_1 + 2p_2 + p_4 + p_5 + 2p_6 + 3p_8$
$M_{201} = 4p_1 + 2p_2 + p_5 + 2p_6 + 3p_8 + p_{11}$
$M_{202} = 4p_1 + 2p_2 + 2p_5 + 2p_6 + 2p_8$

Figure 6.2 Reachability graph of the DP-net in Figure 6.1.

Then, a method to efficiently identify and eliminate redundant constraints in an LMILP is developed. Finally, the whole iterative deadlock prevention policy with the redundancy reduction is given.

6.3.1 Multiobjective Supervisory Control

An LMILP is an ILP problem with s objectives that are defined as follows:

$$lex \ \min\{O_1(x), O_2(x), \dots, O_s(x)\}$$

subject to

$$Ax \geqslant b$$

$$x \geqslant 0$$

where $x \in \mathbb{Z}^n, A \in \mathbb{Z}^{m \times n}, b \in \mathbb{Z}^m$, and $O_i(x) \colon \mathbb{Z}^n \to \mathbb{Z}$, for $i = 1, \dots, s$. The lexicographic method assumes that the objectives are ranked in order of importance. In general, we assume that $O_1(x)$ is the most important and $O_r(x)$ is the least important to a decision maker. In the method, we first obtain the minimal value of $O_1(x)$, called $O_1^*(x)$, subject to the constraints. Next, an ILP is solved with an objective to minimize $O_2(x)$ under the above constraints and $O_1(x) \leqslant O_1^*(x)$. The process is continued until all objectives have been achieved.

6.3.2 Design of an Optimal Control Place

Using the method in Chapter 3, we can design a place invariant to obtain a monitor that forbids a given FBM. In fact, more than one FBM may be forbidden by the place invariant. Let I be a place invariant such that:

$$\sum_{i \in \mathbb{N}_A} l_i \cdot M(p_i) \leqslant \beta \tag{6.1}$$

where l_i $(i \in \mathbb{N}_A)$ and β are the coefficients of I. FBM $M_k \in \mathcal{M}_F^*$ is forbidden by I if

$$\sum_{i \in \mathbb{N}_A} l_i \cdot M_k(p_i) \geqslant \beta + 1. \tag{6.2}$$

A set of variables $f_k \in \{0, 1\}$ $(k \in \mathbb{N}_F^* = \{i | M_i \in \mathcal{M}_F^*\})$ are introduced to represent the relation between place invariant I and FBM $M_k \in \mathcal{M}_F^*$ such that $f_k = 1$ indicates that M_k is forbidden by I and $f_k = 0$ indicates that it is not. Then, constraint (6.2) can be modified as:

$$\sum_{i \in \mathbb{N}_A} l_i \cdot M_k(p_i) \geqslant \beta + 1 - \Gamma \cdot (1 - f_k), \quad \forall M_k \in \mathcal{M}_F^*. \tag{6.3}$$

where Γ is a positive integer constant that is chosen big enough.

To design a structurally simple supervisor, the first objective is to minimize the number of monitors each of which contains a control place and some connected arcs. All markings in $\mathcal{M}_F^*$ must be forbidden by the monitors to prevent the underlying system from entering the deadlock zone. If each monitor forbids as many markings as possible via its designed place invariant, the number of monitors in the supervisor should be small. Thus, the first objective is to minimize

$$O_1 = -\sum_{k \in \mathbb{N}_F^*} f_k. \tag{6.4}$$

Then, the number of arcs contained in the supervisor should also be minimized. Two sets of auxiliary variables are introduced. The first set is $u_n \in \{0, 1\}$ $(n \in \mathbb{N}_T = \{i | t_i \in T\})$ where each element represents whether or not an arc is added from the control place p_c to a transition $t_n \in T$ in the controlled net. Let $[N_c](p_c, \cdot)$ be the incidence vector of p_c in $[N_c]$. Its entry is $[N_c](p_c, n) = W(t_n, p_c) - W(p_c, t_n)$. According to (3.6), $[N_c](p_c, n) = -\sum_{i \in \mathbb{N}_A} l_i \cdot [N_p](i, n)$. Then, we have

$$-\sum_{i \in \mathbb{N}_A} l_i \cdot [N_p](i, n) \geqslant -\Gamma \cdot u_n, \quad \forall n \in \mathbb{N}_T \tag{6.5}$$

Constraints (6.5) imply that if $[N_c](p_c, n) \leqslant -1$, then $u_n = 1$, which indicates an arc from p_c to t_n in the controlled net. Similarly, we introduce another set of variables $v_n \in \{0, 1\}$ in which each element represents whether or not there exists an arc from $t_n \in T$ to p_c. Then, we have

$$\sum_{i\in\mathbb{N}_A} l_i \cdot [N_p](i, n) \geqslant -\Gamma \cdot v_n, \quad \forall n \in \mathbb{N}_T. \tag{6.6}$$

Constraints (6.6) indicate that if $[N_c](p_c, n) \geqslant 1$, then $v_n = 1$, which implies the existence of an arc from t_n to p_c in the controlled net. Thus, to minimize the number of added arcs in the controlled net, we seek to minimize

$$O_2 = \sum_{n\in\mathbb{N}_T} (u_n + v_n). \tag{6.7}$$

Finally, the place invariant coefficients l_i's are minimized to get the simplest expression of the supervisor. Since the place invariant satisfies

$$\sum_{i\in\mathbb{N}_A} l_i \cdot M(p_i) \leqslant \beta, \tag{6.8}$$

we simply seek to minimize

$$O_3 = \beta. \tag{6.9}$$

Note that the coefficients l_i's ($i \in \mathbb{N}_A$) of the place invariant are non-negative. Combining all the above objectives and constraints, we have the following LMILP:

lex $\min\{O_1, O_2, O_3\}$

subject to

$$\sum_{i\in\mathbb{N}_A} l_i \cdot M_m(p_i) \leqslant \beta, \quad \forall M_m \in \mathcal{M}_L^* \tag{6.10}$$

$$\sum_{i\in\mathbb{N}_A} l_i \cdot M_k(p_i) \geqslant \beta + 1 - \Gamma \cdot (1 - f_k), \quad \forall M_k \in \mathcal{M}_F^* \tag{6.11}$$

$$-\sum_{i\in\mathbb{N}_A} l_i \cdot [N_p](i, n) \geqslant -\Gamma \cdot u_n, \quad \forall n \in \mathbb{N}_T \tag{6.12}$$

$$-\sum_{i\in\mathbb{N}_A} l_i \cdot [N_p](i, n) \leqslant \Gamma \cdot v_n, \quad \forall n \in \mathbb{N}_T \tag{6.13}$$

$$l_i \in \{0, 1, 2, \ldots\}, \quad \forall i \in \mathbb{N}_A$$

$$f_k \in \{0, 1\}, \quad \forall k \in \mathbb{N}_F^*$$

$$\beta \in \{1, 2, \ldots\}$$

$$u_n, v_n \in \{0, 1\}, \quad \forall n \in \mathbb{N}_T$$

The objectives in LMILP indicate that a monitor is designed at each iteration to forbid as many markings in $\mathcal{M}_F^*$ as possible by its corresponding place invariant, have the fewest added arcs, and have the simplest coefficients. The numbers of constraints and variables in the LMILP are $|\mathcal{M}_L^*| + |\mathcal{M}_F^*| + 2|T|$ and $|P_A| + |\mathcal{M}_F^*| + 2|T| + 1$, respectively.

6.3.3 Identification of Redundant Constraints

This section presents an approach to identify and eliminate redundant reachability constraints in the above LMILP to further accelerate its solutiion process. In LMILP, Constraints (6.10)–(6.13) determine its feasible region. After removing the objectives of LMILP, we obtain the following system denoted by Y.

System Y :

$$\sum_{i \in \mathbb{N}_A} l_i \cdot M_m(p_i) \leqslant \beta, \quad \forall M_m \in \mathcal{M}_L^*$$

$$\sum_{i \in \mathbb{N}_A} l_i \cdot M_k(p_i) \geqslant \beta + 1 - \Gamma \cdot (1 - f_k), \quad \forall M_k \in \mathcal{M}_F^*$$

$$-\sum_{i \in \mathbb{N}_A} l_i \cdot [N_p](i, n) \geqslant -\Gamma \cdot u_n, \quad \forall n \in \mathbb{N}_T$$

$$-\sum_{i \in \mathbb{N}_A} l_i \cdot [N_p](i, n) \leqslant \Gamma \cdot v_n, \quad \forall n \in \mathbb{N}_T$$

$$l_i \in \{0, 1, 2, \ldots\}, \quad \forall i \in \mathbb{N}_A$$

$$f_k \in \{0, 1\}, \quad \forall k \in \mathbb{N}_F^*$$

$$\beta \in \{1, 2, \ldots\}$$

$$u_n, v_n \in \{0, 1\}, \quad \forall n \in \mathbb{N}_T.$$

To define a redundant reachability constraint in the system, we use $\Omega(Y)$ to denote all possible supervisors satisfying Y and $\tilde{Y}^{<m>}$ to denote the set of constraints obtained by deleting the reachability constraint $\sum_{i \in \mathbb{N}_A} l_i \cdot M_m(p_i) \leqslant \beta$ from Y.

Theorem 6.1 *A reachability constraint*

$$\sum_{i \in \mathbb{N}_A} l_i \cdot M_m(p_i) \leqslant \beta \tag{6.14}$$

is redundant in Y iff

$$\Omega(Y) = \Omega(\tilde{Y}^{<m>}). \tag{6.15}$$

Theorem 6.1 means that a reachability constraint is redundant if and only if the feasible region of Y remains unchanged after the constraint is removed. In addition, Condition (6.15) implies that the following equations should be satisfied:

$$\Omega(Y) \subseteq \Omega(\tilde{Y}^{<m>}) \tag{6.16}$$

$$\Omega(Y) \supseteq \Omega(\tilde{Y}^{<m>}). \tag{6.17}$$

Condition (6.16) is satisfied since the removal of a constraint expands the feasible region. Thus, (6.17) is the center of our focus. In the following, a sufficient and necessary condition for a reachability constraint to be redundant in Y is given in the form of an ILP problem.

Theorem 6.2 *Consider the following ILP problem*

$$\mathbb{P}_m^1 = \max \left\{ \sum_{i \in \mathbb{N}_A} l_i \cdot M_m(p_i) - \beta \right\}$$

subject to

$$\sum_{i \in \mathbb{N}_A} l_i \cdot M_m(p_i) \leqslant \beta, \quad \forall M_m \in \mathcal{M}_L^* \text{ and } l \neq m$$

$$\sum_{i \in \mathbb{N}_A} l_i \cdot M_k(p_i) \geqslant \beta + 1 - \Gamma \cdot (1 - f_k), \quad \forall M_k \in \mathcal{M}_F^*$$

$$-\sum_{i \in \mathbb{N}_A} l_i \cdot [N_p](i, n) \geqslant -\Gamma \cdot u_n, \quad \forall n \in \mathbb{N}_T$$

$$-\sum_{i \in \mathbb{N}_A} l_i \cdot [N_p](i, n) \leqslant \Gamma \cdot v_n, \quad \forall n \in \mathbb{N}_T. \tag{6.18}$$

The reachability constraint expressed in (6.14) is redundant in Y iff $\mathbb{P}_m^1 \leqslant 0$.

Proof: First, we prove that if $\mathbb{P}_m^1 \leqslant 0$, (6.14) is redundant in Y. We know that all feasible solutions of (6.18) constitute $\Omega(\tilde{Y}^{<m>})$. If $\mathbb{P}_m^1 \leqslant 0$ for any solution in $\Omega(\tilde{Y}^{<m>})$, we have:

$$\sum_{i \in \mathbb{N}_A} l_i \cdot M_m(p_i) - \beta \leqslant \mathbb{P}_m^1 \leqslant 0 \tag{6.19}$$

which is the same as the eliminated constraint. Thus, any feasible solution in $\Omega(\tilde{Y}^{<m>})$ is also feasible to system Y. Therefore, (6.17) holds. According to Theorem 6.1, (6.14) is redundant in Y.

Then, we prove that if (6.14) is a redundant constraint, $\mathbb{P}_m^1 \leqslant 0$ holds. If the constraint in (6.14) is redundant, (6.17) should be true, namely, any solution in $\Omega(\tilde{Y}^{<m>})$ is a solution to Y. Since (6.14) should be satisfied by any solution to Y, the maximal value of the objective function, which is the difference between the left side and the right side of (6.14), should be less than 0. Therefore, $\mathbb{P}_m^1 \leqslant 0$ holds. ∎

Theorem 6.2 gives a sufficient and necessary condition for a reachability constraint to be redundant in Y. Problem $\mathbb{P}_m^1$ has $|\mathcal{M}_L^*| + |\mathcal{M}_F^*| + 2|\mathbb{N}_T| - 1$ constraints and $|\mathbb{N}_A| + |\mathcal{M}_F^*| + 2|\mathbb{N}_T| + 1$ variables. Algorithm 6.1 presents a redundancy identification method based on another ILP problem $\mathbb{P}_m^2$, which is revised from $\mathbb{P}_m^1$ but has fewer constraints and variables.

$$\mathbb{P}_m^2 = \max \left\{ \sum_{i \in \mathbb{N}_A} l_i \cdot M_m(p_i) - \beta \right\}$$

subject to Q

where Q is given in Algorithm 6.1.

Algorithm 6.1

```
Redundancy identification algorithm for the LMILP.
Input: The LMILP of a DP-net.
Output: Redundant reachability constraints found in the LMILP.
Redundancy_identification {
        R := ∅; /* R denotes the set of redundant constraints found
in the LMILP. */
        for each reachability constraint γₘ ≤ β
        {
                Q := C_R; /* C_R denotes all reachability constraints. */
                Q := Q̃<m>; /* Q̃<m> is obtained by eliminating γₘ ≤ β
from Q. */
                for each γₗ ≤ β in R
                        Q := Q̃<l>;
                Solve the following problem
                        Pₘ² = max { Σ lᵢ · Mₘ(pᵢ) - β }
                                  i∈N_A
                        subject to Q;
                if Pₘ² ≤ 0
                        R := R ∪ {γₘ ≤ β};
        }
        return R.
}
```

In Algorithm 6.1, $\gamma_m \leq \beta$ is used to denote the reachability constraint $\sum_{i \in \mathbb{N}_A} l_i \cdot M_m(p_i) \leq \beta$ for conciseness. The algorithm can identify redundant reachability constraints in LMILP by using $\mathbb{P}_m^2$. For each reachability constraint $\gamma_m \leq \beta$, $\mathbb{P}_m^2$ is used to judge whether the constraint is redundant or not. First, it selects all the reachability constraints in the LMILP as the candidates Q to be handled. Next, $\gamma_m \leq \beta$ and all redundant constraints found so far are removed from Q. Then, ILP $\mathbb{P}_m^2$ with constraints Q is solved. If the obtained result is less than or equal to 0, the reachability constraint $\gamma_m \leq \beta$ is redundant and added to the set $\mathcal{R}$. The advantage of Algorithm 6.1 is that the numbers of constraints and variables in $\mathbb{P}_m^2$ are at most $|\mathcal{M}_L^*| - 1$ and $|\mathbb{N}_A| + 1$, respectively, both of which are much smaller than those of $\mathbb{P}_m^1$. In addition, all constraints found by Algorithm 6.1 are redundant in Y.

Theorem 6.3 *All constraints of $\mathcal{R}$ obtained by Algorithm 6.1 are redundant in Y.*

Proof: *For any constraint* $\gamma_m \leqslant \beta$ *in* $\mathcal{R}$, *we have* $\mathbb{P}_m^2 \leqslant 0$. *In problem* $\mathbb{P}_m^2$, *the redundant constraints that are found so far are taken away from Q. Therefore, the feasible region of* $\mathbb{P}_m^2$ *is a superset of that of* $\mathbb{P}_m^1$. *Thus, we have* $\mathbb{P}_m^1 \leqslant \mathbb{P}_m^2$, *which implies that* $\mathbb{P}_m^1 \leqslant 0$. *Therefore,* $\gamma_m \leqslant \beta$ *is redundant in system Y according to Theorem 6.2.* ∎

6.3.4 Iterative Deadlock Prevention

This section presents an iterative deadlock prevention method with redundancy reduction to obtain optimal and structurally simple supervisors for DP-nets. It is realized by Algorithm 6.2.

Algorithm 6.2

```
Iterative Deadlock prevention by using the LMILP and redundancy
reduction.
Input: A DP-net N with M₀.
Output: A controlled Petri net with an optimal and simple
supervisor.
Iterative_deadlock_prevention {
        Generate G(N, M₀) and compute 𝓜_F and 𝓜_L;
        Compute 𝓜*_F and 𝓜*_L by the method in Section 3.3.3;
        V_P := ∅, V_A := ∅;  /* V_P and V_A denote the sets of control
places and arcs in the supervisor, respectively. */
        Compute the set 𝓡 of redundant constraints by the method in
Section 6.3.3;
        j := 0;  /* j denotes the iteration number. */
        𝓜'*_F = 𝓜*_F;
        while 𝓜'*_F ≠ ∅
        {
            j := j + 1;
            Design an LMILP by the method in Section~6.3.2;
            Delete 𝓡 from the LMILP and solve the reduced LMILP to
obtain l_i(i ∈ ℕ_A) and β of the underlying place invariant I_j. If it has
no nonzero solution, exit;
            Design a control place p_cⱼ with its arcs A_j by the method
in Section 3.3.2;
            V_P := V_P ∪ {p_cⱼ}, V_A := V_A ∪ {A_j} and 𝓜'*_F := 𝓜'*_F \ F_j;  /* F_j
denotes the set of markings in 𝓜*_F that are forbidden by I_j at the
jth iteration. */
        }
        Add V_P and V_A to (N, M₀) and output the controlled net;
}
```

First, Algorithm 6.2 computes all markings in $\mathcal{M}_F$ and $\mathcal{M}_L$ of $G(N, M_0)$. Next, $\mathcal{M}_F^*$ and $\mathcal{M}_L^*$ are derived by using the vector covering method. Then, redundant reachability constraints are identified by Algorithm 6.1. After that, an iterative process is carried out. At each iteration, an LMILP is formulated. After eliminating redundant reachability constraints from the LMILP, a reduced LMILP is solved to obtain a monitor that forbids as many markings in $\mathcal{M}_F^*$ as possible, possesses the fewest related arcs, and has the simplest coefficients in the corresponding place invariant. Then, all markings that are forbidden by the monitor are removed from $\mathcal{M}_F^{\prime*}$. This process is carried out until $\mathcal{M}_F^{\prime*} = \emptyset$. Finally, all the obtained monitors are added to the original net to obtain a controlled net. Since all FBMs are forbidden by the monitors, the controlled net cannot enter the deadlock zone of the reachability graph. In addition, it is guaranteed that Algorithm 6.2 obtains an optimal supervisor if a nonzero solution exists for LMILP at each iteration.

Theorem 6.4 *Algorithm 6.2 obtains an optimal supervisor for a DP-net if LMILP at each iteration has a nonzero solution.*

Proof: *If LMILP at each iteration has a nonzero solution, at least one marking in $\mathcal{M}_F^*$ is removed from $\mathcal{M}_F^*$ at each iteration. Thus, the algorithm can terminate with a supervisor that forbids all markings in $\mathcal{M}_F^*$. Therefore, all FBMs are forbidden by the obtained supervisor, implying that the controlled system never enters the deadlock zone. On the other hand, if LMILP at each iteration has a nonzero solution, constraints (6.10) are satisfied. That is to say, no legal markings are forbidden by the designed place invariant and an optimal monitor can be obtained at each iteration. Thus, when the iterative process terminates, no legal markings are forbidden by the supervisor that comprises all the obtained optimal monitors. Therefore, if LMILP has a nonzero solution at each iteration, the supervisor obtained by Algorithm 6.2 is optimal. Note that when an iteration of LMILP has a zero solution, the supervisor synthesis algorithm exits with a failure.* ■

We use the DP-net shown in Figure 6.1 to illustrate the presented approach. Its reachability graph has 203 reachable markings of which 192 are legal ones and 11 are FBMs. By using the vector covering method, the minimal covering set of legal markings and the minimal covered set of FBMs have 19 and 3 elements, respectively, i.e., $\mathcal{M}_L^* = \{2p_2 + 2p_3 + 2p_4, 2p_2 + 2p_3 + p_4 + p_5, 2p_2 + p_3 + 2p_5, 2p_2 + p_3 + p_4 + p_5 + p_6, 2p_2 + 2p_5 + p_6, p_2 + p_3 + p_4 + p_5 + p_6 + p_7, p_2 + p_3 + 2p_5 + p_6 + p_7, 2p_5 + 2p_6 + 2p_7, p_2 + 2p_5 + 2p_6 + p_7, p_2 + p_3 + 2p_4 + p_6 + p_7, 2p_2 + p_3 + 2p_4 + p_6, p_2 + 2p_3 + 2p_4 + p_7, p_2 + 2p_3 + p_4 + p_5 + p_7, p_3 + 2p_4 + 2p_7, 2p_4 + p_6 + 2p_7, p_3 + p_4 + p_5 + p_6 + 2p_7, p_4 + p_5 + 2p_6 + 2p_7, p_2 + p_4 + p_5 + 2p_6 + p_7, p_3 + 2p_5 + p_6 + 2p_7\}$ and $\mathcal{M}_F^* = \{2p_2 + p_3 + 2p_5 + p_6, 2p_3 + 2p_5, 2p_2 + 2p_6\}$. Note that only activity places are considered in the supervisor synthesis.

In the first iteration, we generate the reachability constraints (6.10) in LMILP and use Algorithm 6.1 to identify the redundant reachability constraints. The results of Algorithm 6.1 show that $l_1 + l_2 + l_3 + l_4 + l_5 + l_6 \leqslant \beta$ is a redundant. So, $\mathcal{R} = \{l_1 + l_2 + l_3 + l_4 + l_5 + l_6 \leqslant \beta\}$. After deleting the redundant constraint in $\mathcal{R}$, we have the following set of reachability constraints:

$$2l_2 + 2l_3 + 2l_4 \leqslant \beta$$
$$2l_2 + 2l_3 + l_4 + l_5 \leqslant \beta$$
$$2l_2 + l_3 + 2l_5 \leqslant \beta$$
$$2l_2 + l_3 + l_4 + l_5 + l_6 \leqslant \beta$$
$$2l_2 + 2l_5 + l_6 \leqslant \beta$$
$$l_2 + l_3 + 2l_5 + l_6 + l_7 \leqslant \beta$$
$$2l_5 + 2l_6 + 2l_7 \leqslant \beta$$
$$l_2 + 2l_5 + 2l_6 + l_7 \leqslant \beta$$
$$l_2 + l_3 + 2l_4 + l_6 + l_7 \leqslant \beta$$
$$2l_2 + l_3 + 2l_4 + l_6 \leqslant \beta$$
$$l_2 + 2l_3 + 2l_4 + l_7 \leqslant \beta$$
$$l_2 + 2l_3 + l_4 + l_5 + l_7 \leqslant \beta$$
$$l_3 + 2l_4 + 2l_7 \leqslant \beta$$
$$2l_4 + l_6 + 2l_7 \leqslant \beta$$
$$l_3 + l_4 + l_5 + l_6 + 2l_7 \leqslant \beta$$
$$l_4 + l_5 + 2l_6 + 2l_7 \leqslant \beta$$
$$l_2 + l_4 + l_5 + 2l_6 + l_7 \leqslant \beta$$
$$l_3 + 2l_5 + l_6 + 2l_7 \leqslant \beta \tag{6.20}$$

Let I_1 be the place invariant designed at the first iteration. We introduce variables f_1, f_2, and f_3 ($f_1, f_2, f_3 \in \{0, 1\}$) to represent whether or not I_1 forbids $\text{FBM}_1 = 2p_2 + p_3 + 2p_5 + p_6$, $\text{FBM}_2 = 2p_3 + 2p_5$, and $\text{FBM}_3 = 2p_2 + 2p_6$, respectively. Then, we have the following constraints.

$$2l_2 + l_3 + 2l_5 + l_6 \geqslant \beta + 1 - \Gamma \cdot (1 - f_1)$$
$$2l_3 + 2l_5 \geqslant \beta + 1 - \Gamma \cdot (1 - f_2)$$
$$2l_2 + 2l_6 \geqslant \beta + 1 - \Gamma \cdot (1 - f_3) \tag{6.21}$$

Let p_{c_1} be the control place related to I_1. Two sets of variables u_n and v_n ($n \in \{1, 2, \ldots, 8\}$) are introduced to represent whether or not an arc exists from p_{c_1} to t_n and an arc exists from t_n to p_{c_1}, respectively. Then, we have

$$-l_1 \geqslant -\Gamma \cdot u_1$$
$$l_1 - l_2 \geqslant -\Gamma \cdot u_2$$
$$l_2 - l_3 \geqslant -\Gamma \cdot u_3$$
$$l_3 \geqslant -\Gamma \cdot u_4$$

$$-l_4 \geqslant -\Gamma \cdot u_5$$
$$l_4 - l_5 \geqslant -\Gamma \cdot u_6$$
$$l_5 - l_6 \geqslant -\Gamma \cdot u_7$$
$$l_6 \geqslant -\Gamma \cdot u_8$$
$$-l_1 \leqslant \Gamma \cdot v_1$$
$$l_1 - l_2 \leqslant \Gamma \cdot v_2$$
$$l_2 - l_3 \leqslant \Gamma \cdot v_3$$
$$l_3 \leqslant \Gamma \cdot v_4$$
$$-l_4 \leqslant \Gamma \cdot v_5$$
$$l_4 - l_5 \leqslant \Gamma \cdot v_6$$
$$l_5 - l_6 \leqslant \Gamma \cdot v_7$$
$$l_6 \leqslant \Gamma \cdot v_8 \tag{6.22}$$

Finally, the reduced LMILP in the first iteration is given below:

$$lex \min \left\{ -\sum_{k=1}^{3} f_k, \sum_{n=1}^{8} (u_n + v_n), \beta \right\}$$

subject to

(6.20), (6.21), and (6.22)

This LMILP has an optimal solution with $l_1 = l_2 = 3$, $l_4 = 1$, $l_5 = 5$, $\beta = 15$, $f_1 = f_3 = 1$, $u_1 = u_5 = u_6 = v_3 = v_7 = 1$, and all other variables equalling zero. The designed place invariant is I_1 : $3p_2 + 3p_3 + p_5 + 5p_6 + p_{c_1} = 15$. As a result, the added monitor is $M_0(p_{c_1}) = 15$, ${}^{\bullet}p_{c_1} = \{3t_3, 5t_7\}$, and $p_{c_1}^{\bullet} = \{3t_1, t_5, 4t_6\}$. FBM$_1$ and FBM$_3$ are forbidden by I_1, i.e., $F_1 = \{\text{FBM}_1, \text{FBM}_3\}$. Thus, we have $\mathcal{M}_F^{\prime *} = \mathcal{M}_F^{\prime *} \backslash F_1 = \{\text{FBM}_2\}$.

Let I_2 be the designed place invariant at the second iteration. Similarly, we have the reduced LMILP for the second iteration as follows:

$$lex \min \left\{ -\sum_{k=1}^{3} f_k, \sum_{n=1}^{8} (u_n + v_n), \beta \right\}$$

subject to

$$2l_3 + 2l_5 \geqslant \beta + 1 - \Gamma \cdot (1 - f_2),$$

(6.20), and (6.22)

This LMILP has an optimal solution with $l_2 = l_4 = 1$, $\beta = 3$, $f_2 = 1$, $u_2 = u_5 = v_3 = v_6 = 1$, and all other variables equalling zero. The designed place invariant is I_2 : $p_3 + p_5 + p_{c_2} = 3$. Its corresponding monitor is $M_0(p_{c_2}) = 3$, ${}^{\bullet}p_{c_2} = \{t_3, t_6\}$, and $p^\bullet_{c_2} = \{t_2, t_5\}$. FBM$_2$ is forbidden by I_2, namely, $\mathcal{F}_2 = \{\text{FBM}_2\}$. Since $\mathcal{M}'^*_F = \mathcal{M}'^*_F \backslash \mathcal{F}_2 = \emptyset$, the algorithm terminates.

Therefore, the final supervisor has two monitors that contain two control places and nine connected arcs. The detailed results are shown in Table 6.1 where $|\mathcal{R}|$ denotes the number of the found redundant reachability constraints. Its first column shows iteration j, second column gives the designed place invariant I_j, third column, denoted by $|\mathcal{F}_j|$, is the number of markings in $\mathcal{M}^*_F$ that are forbidden by I_j at the jth iteration, the fourth to sixth columns indicate the pre-transitions $^\bullet p_{c_j}$, post-transitions $p^\bullet_{c_j}$, and initial marking $M_0(p_{c_j})$ of the control place p_{c_j}, respectively, the seventh and eighth columns, denoted by N_{var} and N_{con}, are the numbers of variants and constraints in LMILP before redundancy elimination, respectively, and the last column, denoted by N'_{con}, is the number of constraints in the reduced LMILP after redundancy elimination. Adding these two control places and nine arcs to the original net, we obtain a live and maximally permissive controlled net as shown in Figure 6.3 where all 192 legal markings of the original net survive.

Table 6.1 The obtained supervisor for the net shown in Figure 6.1 ($|\mathcal{R}| = 1$)

j	I_j	$\|\mathcal{F}_j\|$	$^\bullet p_{c_j}$	$p^\bullet_{c_j}$	$M_0(p_{c_j})$	N_{var}	N_{con}	N'_{con}
1	$3p_2 + 3p_3 + p_5 + 5p_6 \leqslant 15$	2	$3t_3, 5t_7$	$3t_1, t_5, 4t_6$	15	26	38	37
2	$p_3 + p_5 \leqslant 3$	1	t_3, t_6	t_2, t_5	3	24	36	35

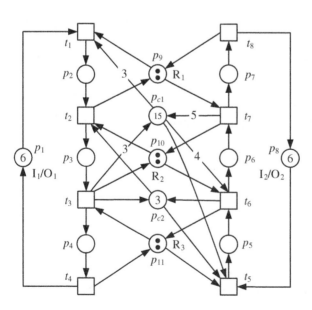

Figure 6.3 Controlled net with a supervisor for the model in Figure 6.1.

6.4   Illustrative Examples

In this section, several resource allocation systems available in the literature are tested to demonstrate the method presented in this chapter. First, the integrated net analyzer (INA) is used to compute $G(N, M_0)$ of a Petri net model N and C++ programs are developed to generate $\mathcal{M}_L^*$ and $\mathcal{M}_F^*$ and eliminate redundant reachability constraints at each iteration. Then, Lingo is used as a linear programming problem solver to find an optimal solution for the reduced LMILP at each iteration. Finally, the behavioral permissiveness of the final results is verified via INA even though our theory guarantees it.

First, consider an automated manufacturing system (Li *et al.*, 2008*a*; Piroddi *et al.*, 2009; Uzam, 2002) that consists of two robots R_1–R_2, four machines M_1–M_4, two loading buffers I_1–I_2, and two unloading buffers O_1–O_2. Two types of parts, P_1 and P_2, are processed in the system. The production sequences are:

$$P_1: I_1 \rightarrow R_1 \rightarrow M_1(\text{or } M_2) \rightarrow R_1 \rightarrow M_3 \rightarrow R_2 \rightarrow O_1$$
$$P_2: I_2 \rightarrow R_2 \rightarrow M_4 \rightarrow R_1 \rightarrow M_2 \rightarrow R_1 \rightarrow O_2$$

Its DP-net is shown in Figure 6.4 that has 19 places and 14 transitions. Its places are divided into idle places $P^0 = \{p_1, p_8\}$, resource places $P_R = \{p_{14} - p_{19}\}$, and activity places $P_A = \{p_2 - p_7, p_9 - p_{13}\}$. It reachability graph $G(N, M_0)$ has 282 markings. Among them 205 are legal markings and 54 are FBMs. By using the vector covering method, $\mathcal{M}_L^*$ and $\mathcal{M}_F^*$ have 26 and 8 markings, respectively. By using Algorithm 6.2, we obtain the results as shown in Table 6.2 where $\mathcal{T}_R$ denotes the computational time for redundancy identification at each iteration, $\mathcal{T}_{LP}$ represents the solution time of the original LMILP at each iteration, and $\mathcal{T}_{LP}'$ is the solution time of the reduced LMILP at each iteration. For this system, there is no redundant reachability constraint. The redundancy identification takes 0.09 seconds and the iterative process of Algorithm 6.2 is completed within one second.

This benchmark system has been widely used and tested in the literature. Table 6.3 shows the available results on the supervisor synthesis for the system regarding the type of method ("1" denotes the structural analysis method and "2" represents the reachability graph analysis method), the numbers of control places, added arcs, and reachable markings of the controlled net. Note that in Chen *et al.* (2012), only the results of MFFP1 are used for comparisons since the efficiency of its another method, MFFP2, greatly depends on the manual selection of $M_j \in \mathcal{M}_F^*$ to compute a place invariant at each iteration. From Table 6.3, we see that all methods including Algorithm 6.2 can obtain the same optimal supervisors. However, Algorithm 6.2 significantly outperforms the structural analysis methods (Li *et al.*, 2008*a*; Piroddi *et al.*, 2008; Zhang & Wu, 2013) in terms of the numbers of added places and arcs. Among the

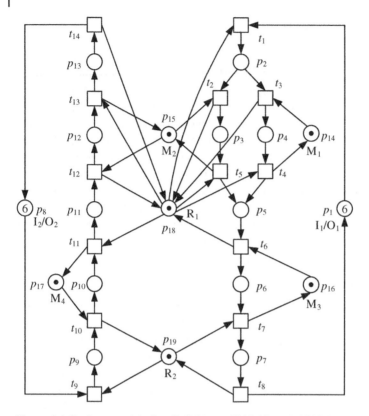

Figure 6.4 Petri net model of an AMS (Uzam, 2002; Li *et al.*, 2008*a*).

reachability graph analysis methods, the supervisor obtained by Algorithm 6.2 has fewer places and arcs than those of Chen *et al.* (2011, 2012) and Uzam (2002). It has two more arcs than those in Chen and Li (2011) and Huang *et al.* (2015*c*), but it is more computationally efficient than them. The MMP in Chen and Li (2011) has 328 constraints and 152 variables and the solution time is about two seconds. The LMPA in Huang *et al.* (2015*c*) has 552 constraints and 376 variables and its solution needs 45 seconds. But Algorithm 6.2 has two ILPs that are much fewer constraints and variables than those in Chen and Li (2011) and Huang *et al.* (2015*c*) and the whole algorithm terminates within one second.

Then, we test a more complex system in Ezpeleta *et al.* (1995) that is also a benchmark system widely used in the literature. It has three robots R_1–R_3, four kinds of machines M_1–M_4, three loading buffers I_1–I_3, and three unloading buffers O_1–O_3.

Table 6.2 Added places and arcs for the net shown in Figure 6.4. ($|\mathcal{R}| = 0$, $\mathcal{T}_{\mathcal{R}} = 0.09\ s$)

j	l_j	$|\mathcal{F}_j|$	$\bullet p_{c_j}$	$p^\bullet_{c_j}$	$M_0(p_{c_j})$	N_{var}	N_{con}	$\mathcal{T}_{LP}$	N'_{con}	$\mathcal{T}'_{LP}$
1	$4p_2 + 9p_3 + 4p_4 + 5p_5 + p_9 + p_{10} + 8p_{11} + 8p_{12} \leqslant 15$	5	$4t_5, 5t_6, 8t_{13}$	$4t_1, 5t_2, t_4, t_9, 7t_{11}$	15	48	62	1 s	62	1 s
2	$p_2 + 2p_3 + p_4 + 2p_5 + 2p_6 + 3p_9 + 3p_{10} \leqslant 9$	3	$2t_7, 3t_{11}$	$t_1, t_2, t_4, 3t_9$	9	43	57	0 s	57	0 s
							Total	1 s		1 s

Table 6.3 Performance comparisons for the net shown in Figure 6.4.

Parameters	Li et al. (2008a)	Piroddi et al. (2008)	Zhang and Wu (2013)	Chen et al. (2011)	Uzam (2002)	Chen et al. (2012)	Chen and Li (2011)	Huang et al. (2015c)	Algorithm 6.2
Type of method	1	1	1	2	2	2	2	2	2
No. control places	9	5	4	8	6	2	2	2	2
No. added arcs	42	23	18	37	32	15	12	12	14
No. markings	205	205	205	205	205	205	205	205	205

Three types of parts, P_1–P_3, are processed in the system. The production sequences are:

$$P_1: I_1 \rightarrow R_1 \rightarrow M_1 \rightarrow R_2 \rightarrow M_2 \rightarrow R_3 \rightarrow O_1$$
$$\text{or } I_1 \rightarrow R_1 \rightarrow M_3 \rightarrow R_2 \rightarrow M_4 \rightarrow R_3 \rightarrow O_1$$
$$P_2: I_2 \rightarrow R_2 \rightarrow M_2 \rightarrow R_2 \rightarrow O_2$$
$$P_3: I_3 \rightarrow R_3 \rightarrow M_4 \rightarrow R_2 \rightarrow M_3 \rightarrow R_1 \rightarrow O_3$$

Figure 6.5 shows its DP-net model. It has 20 transitions and 26 places that are divided into $P^0 = \{p_1, p_5, p_{14}\}$, $P_R = \{p_{20} - p_{26}\}$, and $P_A = \{p_2 - p_4, p_6 - p_{13},$

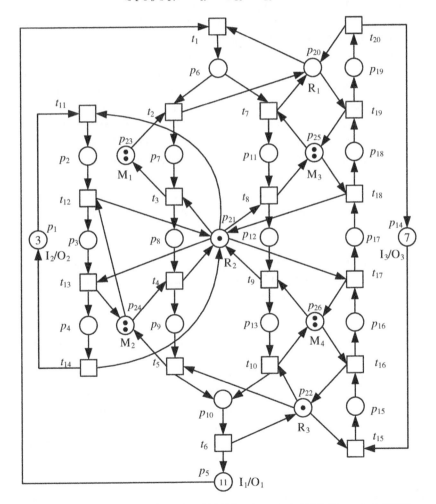

Figure 6.5 Petri net model of another AMS (Ezpeleta *et al.*, 1995; Chen *et al.*, 2011).

$p_{15} - p_{19}$}. Its reachability graph has 26 750 reachable markings, 21 581 of which are legal markings and 4 211 of which are FBMs. By using the vector covering method, $\mathcal{M}_L^*$ and $\mathcal{M}_F^*$ have 393 and 34 markings, respectively. The results of Algorithm 6.2 are shown in Table 6.4. For this system, Algorithm 6.1 costs about 2.78 seconds to identify 47 redundant reachability constraints and the iterative process of Algorithm 6.2 is completed within 8 minutes 19 seconds.

Table 6.5 shows the available results for the system by using different supervisor synthesis methods in the literature. It can be seen that all reachability graph analysis methods including Algorithm 6.2 are maximally or highly permissive since their numbers of reachable markings in the controlled nets are equal or roughly equal to that of the plant net, i.e., 21 581. Among the reachability graph analysis methods, the supervisor obtained by Algorithm 6.2 has fewer control places and added arcs than those in Chen *et al.* (2011, 2012) and Uzam and Zhou (2007). It has one more control place than those in Chen and Li (2011) and Huang *et al.* (2015c) but 13 fewer arcs and one fewer arc than those in Chen and Li (2011) and Huang *et al.* (2015c), respectively. More importantly, the ILPs in Algorithm 6.2 are much smaller and can be solved much faster than those in Chen and Li (2011) and Huang *et al.* (2015c). For instance, the MMP in Chen and Li (2011) has 15 640 constraints and 1 700 variables, and requires about 30 hours to obtain an optimal solution on our computer. The LMPA in Huang *et al.* (2015c) has 17 000 constraints and 3 040 variables, and needs over 300 hours to obtain a result. However, Algorithm 6.2 formulates six ILPs that has much fewer constraints and variables (see Table 6.4) than those in Chen and Li (2011) and Huang *et al.* (2015c). As a result, the total solution time is less than nine minutes, including the time of generating reachability graph (about six seconds), computing $\mathcal{M}_L^*$ and $\mathcal{M}_F^*$ (about 15 seconds), redundancy eliminating (about three seconds), and ILP solving (8 minutes and 19 seconds). For this net, we have the following important conclusions.

1) From the viewpoint of behavioral permissiveness, Algorithm 6.2 can obtain an optimal supervisor.
2) Concerning the structural complexity, Algorithm 6.2 obtains a compressed supervisor with six control places, only obtains one more place than Chen and Li (2011) and Huang *et al.* (2015c), and has 42 arcs, which is minimal among all the optimal supervisors obtained in the literature.
3) The method is efficient. It reduces the ILP solution time from 30 hours of the MMP in Chen and Li (2011) and over 300 hours of the LMPA in Huang *et al.* (2015c) to about 40 minutes by solving much smaller ILPs. In addition, the redundancy reduction process only takes about three seconds and it further reduces the ILP solution time to less than nine minutes.

Table 6.4 Added places and arcs for the net shown in Figure 6.5. ($|R| = 47$, $T_R = 2.78$ s)

j	l_j	$|F_j|$	$^\bullet p_q$	$p_q^\bullet$	$M_0(p_q)$	N_{var}	N_{con}	T_{LP}	N'_{con}	T'_{LP}
1	$33p_2 + 33p_3 + 33p_4 + 6p_6 + 6p_7 + 38p_8 + 38p_9 + 6p_{11} + 104p_{12} + 104p_{13} + 107p_{15} + 107p_{16} + p_{17} + 5p_{18} \leq 421$	19	$38t_5, 104t_{10}, 33t_{14}, 106t_{17}, 5t_{19}$	$6t_1, 32t_3, 98t_8, 33t_{11}, 107t_{15}, 4t_{18}$	421	91	467	37 min 22 s	420	7 min 24 s
2	$p_2 + p_3 + p_8 + p_9 + 27p_{11} + 3p_{12} + 3p_{13} + 6p_{15} + 9p_{16} + 18p_{17} \leq 71$	8	$t_5, 24t_8, 3t_{10}, t_{13}, 18t_{18}$	$t_3, 27t_7, t_{11}, 6t_{15}, 3t_{16}, 9t_{17}$	71	72	448	1 m 48 s	401	41 s
3	$7p_6 + 7p_7 + p_8 + p_9 + p_{10} + 7p_{11} + p_{12} + p_{13} + p_{15} + p_{16} + 9p_{17} + 6p_{18} \leq 41$	3	$6t_3, t_6, 6t_8, 3t_{18}, 6t_{19}$	$7t_1, t_{15}, 8t_{17}$	41	64	440	22 s	393	12 s
4	$p_2 + p_3 + p_8 \leq 2$	2	t_4, t_{13}	t_3, t_{11}	2	61	437	1 s	390	1 s
5	$p_{13} + p_{15} \leq 2$	1	t_{10}, t_{16}	t_9, t_{15}	2	59	435	1 s	388	1 s
6	$p_{12} + p_{16} \leq 2$	1	t_9, t_{17}	t_8, t_{16}	2	58	434	0 s	387	0 s
							Total	39 m 34 s		8 m 19 s

Table 6.5 Performance comparisons for the net shown in Figure 6.5.

Parameters	Ezpeleta et al. (1995)	Li and Zhou (2004)	Huang et al. (2001)	Piroddi et al. (2008)	Zhang and Wu (2013)	Uzam and Zhou (2007)	Chen et al. (2011)	Chen et al. (2012)	Chen and Li (2011)	Huang et al. (2015c)	Algorithm 6.2
Type of method	1	1	1	1	1	2	2	2	2	2	2
No. control places	18	6	16	13	12	19	17	6	5	5	6
No. added arcs	106	32	88	82	68	112	101	45	55	43	42
No. markings	6 287	6 287	12 656	21 581	21 581	21 562	21 581	21 581	21 581	21 581	21 581

6.5   Concluding Remarks

This chapter presents an iterative deadlock prevention method with redundancy reduction that can be used to design optimal supervisors with a simple structure regarding both added places and arcs for DP-nets. The advantages of the method are that i) the supervisor synthesis process is well accelerated by using an iterative method and the redundancy elimination technique without compromising the result's optimality and ii) the obtained supervisor is structurally compressed regarding both control places and related arcs.

The only problem with the method is that it still requires the complete computation of a reachability graph given a DP-net. Thus, the presented method is still inapplicable to the Petri net models whose reachability graph cannot be generated within an acceptable period of time. In Piroddi *et al.* (2008, 2009), the strategies of combining reachability graph and Petri net structure analysis have been developed for the supervisor synthesis. How to utilize the information of Petri net structures to compute all legal markings and FBMs without generating the whole reachability graph for all subclasses of DP-nets warrants future research.

6.6   Bibliographical Notes

Computational speed is an important criterion in evaluating optimal supervisor synthesis policies. Most material reported in this chapter on the synthesis of optimal and structure-simple supervisors comes from Huang *et al.* (2018*a*). Other efficient supervisor synthesis approaches are based on elementary siphons (Li & Zhou, 2008), hybrid strategy (Piroddi *et al.*, 2008, 2009), binary decision graphs (Chen *et al.*, 2011), and branch-and-bound schemes (Cordone & Piroddi, 2013). The well-known supervisor computation on place invariants can be seen in Yamalidou *et al.* (1996). The concept of FBMs is developed in Uzam (2002) and Uzam and Zhou (2007). The classification of places in the Petri net models can be found in Ezpeleta *et al.* (1995), Park and Reveliotis (2001), and Zhou and DiCesare (1991).

7

Supervisor Synthesis with Uncontrollable and Unobservable Transitions

This chapter presents a deadlock prevention method based on DP-nets for resource allocation systems with uncontrollable and unobservable events. First, the concepts of admissible markings and first-met inadmissible markings (FIMs) are introduced. Next, place invariants are designed to survive all admissible markings and prohibit all FIMs to keep the system from reaching deadlock markings, livelock markings, bad markings, and the markings that can evolve into deadlocks, livelocks, or bad markings via the firings of uncontrollable transitions. Then, an integer linear program is developed to ensure that the obtained supervisor does not violate transition unobservability and it is structurally minimal in terms of control places and added arcs. The approach can deal with the nets in which some crucial transitions are uncontrollable and/or unobservable. In addition, the resulting supervisors are admissible, i.e., all admissible markings survive in the controlled net. The condition under which the obtained supervisors are optimal (maximally permissive) is also given. Finally, several benchmark systems adapted from the literature are tested to illustrate the reported approach.

7.1 Introduction

Resource allocation systems (RASs) can concurrently process raw and intermediate objects (such as parts, vehicles, and data) with pre-established operation sequences and limited shared resources (such as machines, robots, drives, and programs). The competition for the limited resources among several parallel processes often leads to deadlocks in which two or more operations are each indefinitely waiting for another to release its acquired resources and thus none can be completed. Deadlocks can result in unnecessary economic costs and even catastrophic failures; so they are highly undesirable phenomena.

Supervisory Control and Scheduling of Resource Allocation Systems: Reachability Graph Perspective,
First Edition. Bo Huang and MengChu Zhou.
© 2020 The Institute of Electrical and Electronics Engineers, Inc. Published 2020 by John Wiley & Sons, Inc.

To handle RAS deadlocks, Petri nets are widely adopted to model RASs and address the deadlock problem since Petri nets are suitable to graphically describe concurrency, conflict, and synchronization, detect behavioral properties, and find optimized scheduling schemes in RASs (Hrúz & Zhou, 2007; Li & Zhou, 2004; Mejía & Niño, 2017). In practice, controllability and observability are two important event properties in RASs and they can be modeled by some kinds of transitions in a Petri net formalism. For example, we can use uncontrollable transitions to represent the events that cannot be prevented from happening by any control agent and use unobservable transitions to denote the events that cannot be measured or detected by any control agent (Giua *et al.*, 1992; Moody & Antsaklis, 2000; Qin *et al.*, 2012). For those with uncontrollable transitions, Ghaffari *et al.* (2003) propose a Petri net supervisor synthesis method based on event separation instances to ensure that the obtained supervisor is behaviorally optimal in the presence of uncontrollable transitions. Wang *et al.* (2016) also propose an optimal supervisor synthesis method for such Petri nets by using a bottom-up approach. In You *et al.* (2017*a*), a supervisor synthesis method for a specific subclass of Petri net with uncontrollable transitions is given. To deal with both uncontrollable and unobservable transitions, Moody and Antsaklis (2000) present an approach that transforms a given generalized mutual exclusion constraint into an admissible one to enforce a conjunction of a set of constraints on reachable markings. Based on it, Luo and Zhou (2017) propose an efficient equivalent transformation method for a class of constraints in the Petri nets with uncontrollability and unobservability.

These methods can obtain behaviorally maximal or highly maximal supervisors for the PNs with uncontrollable and/or unobservable transitions. However, in practice, structural simplicity is also an important criterion in the RAS implementation. A structure-minimal supervisor usually implies the lowest implementation cost of the supervisor. For example, in practical implementation, control places are often implemented by some processing devices such as computers, microcontrollers, and programmable logic controllers, and the connected arcs are usually implemented by sensors or actuators. Chen and Barkaoui (2014) propose an integer linear program (ILP) based method to obtain deadlock-free supervisors for PNs with uncontrollable and unobservable transitions. The optimality and the structural simplification of the obtained supervisors can be guaranteed, however, the method requires that optimal supervisors exist and crucial transitions of the net are controllable.

This chapter presents a method that obtains admissible and structure-minimal supervisors for Petri nets of RASs with uncontrollable and unobservable transitions, including the nets whose optimal supervisors do not exist and/or crucial transitions may be uncontrollable. First, the definitions of admissible markings and FIMs are presented for the DP-nets with uncontrollable and unobservable transitions. Then, place invariants are designed via an ILP to survive all admissible markings and forbid all FIMs. These place invariants guarantee that the controlled system is deadlock-free and always stay in the zone of admissible markings. In addition, the obtained supervisor needs not observe any unobservable transition and is structurally minimal. The admissibility of the obtained supervisors is proven. The condition under which the obtained supervisors are optimal is also given.

7.2 Supervisor Synthesis with Uncontrollability and Unobservability

This section presents the approach to design admissible and structure-minimal supervisors for DP-nets with uncontrollable and/or unobservable transitions. First, uncontrollable and unobservable transitions are introduced into the definitions of DP-nets. Next, admissible markings and FIMs of such nets are defined. Based on these definitions, a place invariant-based method to design an admissible monitor is presented. Finally, the procedures to minimize the structure of a supervisor for such DP-nets by formulating a lexicographic multiobjective ILP are given.

7.2.1 DP-Nets with Uncontrollable and/or Unobservable Transitions

To handle uncontrollable and unobservable events in RASs, the DP-nets in Definition 3.1 should be augmented with the following statements. Transitions T are classified into three disjoint subclasses: controllable transitions T_C, which are controllable and observable transitions, observable but uncontrollable transitions $T_{O\bar{C}}$, and unobservable transitions $T_{\bar{O}}$, which are uncontrollable and unobservable transitions, such that $T = T_C \cup T_{O\bar{C}} \cup T_{\bar{O}}$.

In addition, a transition t is called an uncontrollable transition, denoted as $t \in T_{\bar{C}}$, if the firing of t should not be prevented by any added supervisor in any case; otherwise, $t \in T_C$. A transition is called an unobservable transition, i.e., $t \in T_{\bar{O}}$, if the firing of t should not be measured or detected by any supervisor in any case; otherwise, $t \in T_O$ where T_O denotes the set of observable transitions. A controllable transition is also observable, i.e., $T_C \subseteq T_O$, and an unobservable transition is also uncontrollable, i.e., $T_{\bar{O}} \subseteq T_{\bar{C}}$. Hence, we have $T_O = T_C \cup T_{O\bar{C}}$ and $T_{\bar{C}} = T_{\bar{O}} \cup T_{O\bar{C}}$.

To illustrate it, we consider a flexible manufacturing system adapted from Huang *et al.* (2015c). The system has three shared resources R_1–R_3, two loading

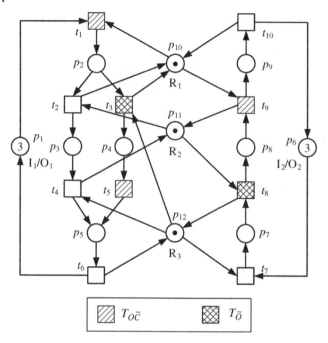

Figure 7.1 A DP-net with $T_{O\tilde{C}} = \{t_1, t_5, t_9\}$ and $T_{\tilde{O}} = \{t_3, t_8\}$.

buffers I_1–I_2, and two unloading buffers O_1–O_2. Two kinds of parts, P_1 and P_2, are processed in the system. Its production sequences are:

$$P_1: I_1 \rightarrow R_1 \rightarrow R_2(\text{or } R_3) \rightarrow R_3 \rightarrow O_1$$
$$P_2: I_2 \rightarrow R_3 \rightarrow R_2 \rightarrow R_1 \rightarrow O_2$$

Its DP-net is shown in Figure 7.1, which has 10 transitions and 12 places with idle places $P_0 = \{p_1, p_6\}$, resource places $P_R = \{p_{10} - p_{12}\}$, and activity places $P_A = \{p_2 - p_5, p_7 - p_9\}$. Suppose that the set of observable but uncontrollable transitions is $T_{O\tilde{C}} = \{t_1, t_5, t_9\}$ and the set of unobservable transitions is $T_{\tilde{O}} = \{t_3, t_8\}$. They are depicted in Figure 7.1 by two different squares. Since an unobservable transition is implicitly uncontrollable, we have $T_{\tilde{C}} = \{t_1, t_3, t_5, t_8, t_9\}$.

7.2.2 Admissible Markings and First-Met Inadmissible Markings

In the theory of regions (Uzam, 2002), the reachability graph of a Petri net is divided into a live zone and a deadlock zone. The live zone contains all legal markings $\mathcal{M}_L$ that can reach the initial marking M_0, and the deadlock zone consists of deadlock markings, livelock markings, and bad markings that inevitably lead to deadlocks and livelocks. The markings in the deadlock zone are denoted as

$\mathcal{M}_D$. If a DP-net N has uncontrollable transitions $T_{\tilde{C}}$, the net has the possibility of uncontrollably evolving into the deadlock zone via some uncontrollable transitions. Thus, to prevent deadlocks from happening in such a net, the designed supervisor should keep the system from reaching not only deadlocks, livelocks, and bad markings but also the markings that evolve into deadlocks, livelocks, or bad markings via the firings of uncontrollable transitions only. Meanwhile, to maximize the behavioral permissiveness of the controlled net, the supervisor should survive as many "good" markings as possible. Note that RASs are human-made systems in which a "good" state permitted by the system often means a state that is not a deadlock and can reach the initial state. Thus, the controlled net of an RAS with uncontrollable events should run in a maximal strongly connected component (SCC) of $G(N, M_0)$ that contains M_0 and should never run outside SCC via the firings of uncontrollable transitions.

Definition 7.1 *Given a DP-net N with uncontrollable and/or unobservable transitions and an initial marking M_0, its reachable markings $R(N, M_0)$ consist of two disjoint classes of markings: admissible markings $\mathcal{M}_A$ and inadmissible markings $\mathcal{M}_{\tilde{A}}$. $\mathcal{M}_A$ is defined as the maximal subset of $R(N, M_0)$ such that (i) $\forall M \in \mathcal{M}_A$, M can reach M_0 and M_0 can reach M along a path without leaving $\mathcal{M}_A$; and (ii) each edge that connects a marking in $\mathcal{M}_A$ to a marking outside $\mathcal{M}_A$ is labeled with a controllable transition.*

It means that, in $G(N, M_0)$, all admissible markings and the edges among them form the maximal SCC that contains the initial marking M_0. In addition, there is no uncontrollable transition whose firing leads the system from an admissible marking to an inadmissible one. Note that the definition of admissible markings in Definition 7.1 is different from that in Moody and Antsaklis (2000) since admissible markings in Definition 7.1 are reachable in both the plant net and the controlled net, while the ones in Moody and Antsaklis (2000) may not be reachable due to either the initial marking of the plant net or the restrictions of the supervisor. Algorithm 7.1 gives the procedures to generate $\mathcal{M}_A$ and $\mathcal{M}_{\tilde{A}}$ for a DP-net with uncontrollable and/or unobservable transitions.

Algorithm 7.1

```
Generation of admissible and inadmissible markings.
Input: G(N, M_0) of a DP-net N with T = T_C ∪ T_{OC̃} ∪ T_{Õ}.
Output: M_A and M_Ã.
Compute_admissible_and_inadmissible_markings {
1        S := G(N, M_0);
2        Compute the maximal SCC of S that contains M_0, denoted
as S'; S := S';
3        Compute the set  of uncontrollable dangerous markings
M_C^D in S, each of which leads the system outside S via the
```

firing of uncontrollable transitions, i.e., $\mathcal{M}_{\widetilde{C}}^{D} := \{M \in S | \exists\ M' \notin S,$
$\sigma \in T_{\widetilde{C}}^{*} \setminus \{\lambda\},\ \text{s.t.}\ M[\sigma\rangle\ M'\}$ where $T_{\widetilde{C}}^{*} \setminus \{\lambda\}$ denotes the set of all finite
strings of elements in $T_{\widetilde{C}}$, excluding the empty string λ.
```
4        If M_C^D ≠ Ø, then remove M_C^D from S and go to Step 2;
5        M_A := S and M_Ā := R(N,M_0) \ M_A;
6        Output M_A and M_Ā.
}
```

In Definition 3.2, optimal supervisors are defined as the ones that can survive all legal markings in the live zone and forbid all illegal markings in the deadlock zone. But they are defined for DP-nets without uncontrollable transitions or unobservable transitions. It may be restrictive for DP-nets with uncontrollable and/or unobservable transitions. Thus, we define admissible supervisors as follows.

Definition 7.2 *For a DP-net with uncontrollable and/or unobservable transitions, a supervisor is said to be admissible if it permits all markings in $\mathcal{M}_A$ and forbids all markings in $\mathcal{M}_{\bar{A}}$.*

Note that, an admissible supervisor is optimal if $\mathcal{M}_L = \mathcal{M}_A$. In addition, after adding an admissible supervisor to a DP-net, the controlled net is reversible since all the permitted markings are in SCC that contains M_0. In the following, we show that, to prevent the net from entering the zone of $\mathcal{M}_{\bar{A}}$, only some markings in $\mathcal{M}_{\bar{A}}$, called FIMs, need to be considered.

Definition 7.3 *In the reachability graph $G(N, M_0)$ of a DP-net N with uncontrollable and/or unobservable transitions, the first inadmissible markings from $\mathcal{M}_A$ to $\mathcal{M}_{\bar{A}}$ are called FIMs, i.e.,*

$$\mathcal{M}_{FIM} = \{M \in \mathcal{M}_{\bar{A}} | \exists M' \in \mathcal{M}_A,\quad t \in T_C,\quad \text{s.t.}\ M'[t\rangle M\}. \tag{7.1}$$

Note that any edge that connects an admissible marking to an FIM is associated with a controllable transition $t \in T_C$. If a supervisor survives all markings in $\mathcal{M}_A$ and forbids all FIMs in $\mathcal{M}_{\bar{A}}$, the controlled net with the supervisor can run in $\mathcal{M}_A$ and never enters $\mathcal{M}_{\bar{A}}$ via the firing of any transition whether it is controllable or not. In such a case, the supervisor is admissible. For the example system given in Figure 7.1, its reachability graph is shown in Figure 7.2 where a compact multiset formalism $\sum_i M(p_i)p_i$ is used to denote a marking M for conciseness. For instance, its initial markings $M_0 = (3,0,0,0,0,3,0,0,0,1,1,1)$ is represented as $3p_1 + 3p_6 + p_{10} + p_{11} + p_{12}$. By using Algorithm 7.1, we have that the set of admissible markings is $\mathcal{M}_A = \{M_0 - M_8, M_{10}, M_{12} - M_{20}, M_{23}, M_{24}, M_{26} - M_{29}\}$ and the rest markings constitute $\mathcal{M}_{\bar{A}}$. In addition, all markings in $\mathcal{M}_{\bar{A}}$ are FIMs, i.e., $\mathcal{M}_{FIM} = \mathcal{M}_{\bar{A}} = \{M_9, M_{11}, M_{21}, M_{22}, M_{25}\}$.

Note that all markings in $\mathcal{M}_A$ and their edges in $G(N, M_0)$ form the maximal SCC that contains M_0. In addition, each edge that leads outside the SCC is controllable.

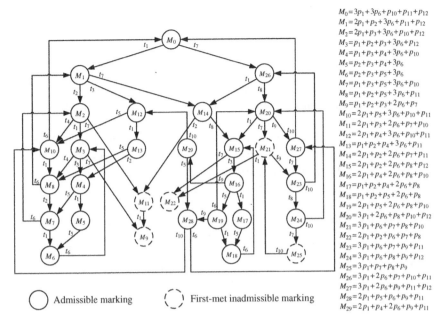

$$M_0 = 3p_1 + 3p_6 + p_{10} + p_{11} + p_{12}$$
$$M_1 = 2p_1 + p_2 + 3p_6 + p_{11} + p_{12}$$
$$M_2 = 2p_1 + p_3 + 3p_6 + p_{10} + p_{12}$$
$$M_3 = p_1 + p_2 + p_3 + 3p_6 + p_{12}$$
$$M_4 = p_1 + p_3 + p_4 + 3p_6 + p_{10}$$
$$M_5 = p_2 + p_3 + p_4 + 3p_6$$
$$M_6 = p_2 + p_3 + p_5 + 3p_6$$
$$M_7 = p_1 + p_3 + p_5 + 3p_6 + p_{10}$$
$$M_8 = p_1 + p_2 + p_5 + 3p_6 + p_{11}$$
$$M_9 = p_1 + p_2 + p_3 + 2p_6 + p_7$$
$$M_{10} = 2p_1 + p_5 + 3p_6 + p_{10} + p_{11}$$
$$M_{11} = 2p_1 + p_3 + 2p_6 + p_7 + p_{10}$$
$$M_{12} = 2p_1 + p_4 + 3p_6 + p_{10} + p_{11}$$
$$M_{13} = p_1 + p_2 + p_4 + 3p_6 + p_{11}$$
$$M_{14} = 2p_1 + p_2 + 2p_6 + p_7 + p_{11}$$
$$M_{15} = 2p_1 + p_2 + 2p_6 + p_8 + p_{12}$$
$$M_{16} = 2p_1 + p_4 + 2p_6 + p_8 + p_{10}$$
$$M_{17} = p_1 + p_2 + p_4 + 2p_6 + p_8$$
$$M_{18} = p_1 + p_2 + p_5 + 2p_6 + p_8$$
$$M_{19} = 2p_1 + p_5 + 2p_6 + p_8 + p_{10}$$
$$M_{20} = 3p_1 + 2p_6 + p_8 + p_{10} + p_{12}$$
$$M_{21} = 3p_1 + p_6 + p_7 + p_8 + p_{10}$$
$$M_{22} = 2p_1 + p_2 + p_6 + p_7 + p_8$$
$$M_{23} = 3p_1 + p_6 + p_7 + p_9 + p_{11}$$
$$M_{24} = 3p_1 + p_6 + p_8 + p_9 + p_{12}$$
$$M_{25} = 3p_1 + p_7 + p_8 + p_9$$
$$M_{26} = 3p_1 + 2p_6 + p_7 + p_{10} + p_{11}$$
$$M_{27} = 3p_1 + 2p_6 + p_9 + p_{11} + p_{12}$$
$$M_{28} = 2p_1 + p_5 + p_6 + p_9 + p_{11}$$
$$M_{29} = 2p_1 + p_4 + 2p_6 + p_9 + p_{11}$$

○ Admissible marking   (⌐) First-met inadmissible marking

Figure 7.2 Reachability graph of the DP-net in Figure 7.1 with $T_{O\hat{C}} = \{t_1, t_5, t_9\}$ and $T_{\hat{O}} = \{t_3, t_8\}$.

Thus, if all markings in $\mathcal{M}_A$ are permitted and all FIMs are forbidden by a designed supervisor, then the controlled net survives all admissible markings and never enters $\mathcal{M}_{\bar{A}}$ controllably or uncontrollably.

7.2.3 Design of an Admissible Monitor

For a DP-net N with $T_{\hat{C}} = T_{O\hat{C}} \cup T_{\hat{O}}$, an admissible and deadlock-free supervisor should prohibit all FIMs and permit all admissible markings in $G(N, M_0)$. A supervisor consists of at least one monitor that comprises a control place and several arcs connected between the place and some transitions of N. To forbid a marking $M' \in \mathcal{M}_{\text{FIM}}$, a monitor should be designed based on a place invariant such that

$$\sum_{i \in \mathbb{N}_A} l(i) \cdot M'(p_i) \geq \beta + 1, \tag{7.2}$$

where $l(i)$ and β are the coefficients of the place invariant and $\mathbb{N}_A = \{i | p_i \in P_A\}$. On the other hand, to survive all admissible markings with the monitor, the following constraint should be satisfied.

$$\sum_{i \in \mathbb{N}_A} l(i) \cdot M(p_i) \leq \beta, \quad \forall M \in \mathcal{M}_A. \tag{7.3}$$

Substituting (7.2) into (7.3), a constraint that permits all admissible markings and forbids M' is obtained as:

$$\sum_{i \in \mathbb{N}_A} l(i) \cdot \left(M(p_i) - M'(p_i)\right) \leq -1, \quad \forall M \in \mathcal{M}_A. \tag{7.4}$$

Constraint (7.4) determines the coefficients l and β of the place invariant. By using the supervisor computation method presented in Section 3.3.2, we can obtain a monitor that consists of a control place and some arcs and forbids an FIM and survives all admissible markings. However, there are usually many FIMs and admissible markings in $G(N, M_0)$ since the size of $G(N, M_0)$ grows exponentially with the net size. To reduce the markings to be considered in such a process, a vector covering method adapted from Section 3.3.3 can be used.

Definition 7.4 *For a DP-net N, $\forall M, M' \in R(N, M_0)$, we say M covers M', denoted as $M \succcurlyeq M'$, if $\forall p \in P_A$, $M(p) \geq M'(p)$.*

If $M \succcurlyeq M'$ and a place invariant I forbids M', i.e., the place invariant I satisfies (7.2), then I also forbids M. Similarly, if $M \succcurlyeq M'$ and a place invariant I permits M, i.e., the place invariant I satisfies (7.3), then I also permits M'.

Definition 7.5 $\mathcal{M}_{FIM}^*$ *is called a minimal covered set of $\mathcal{M}_{FIM}$ if*

1) $\mathcal{M}_{FIM}^* \subseteq \mathcal{M}_{FIM}$;
2) $\forall M \in \mathcal{M}_{FIM}$, $\exists M' \in \mathcal{M}_{FIM}^*$ *such that $M \succcurlyeq M'$; and*
3) $\forall M \in \mathcal{M}_{FIM}^*$, $\nexists M'' \in \mathcal{M}_{FIM}^*$ *such that $M \neq M''$ and $M \succcurlyeq M''$.*

Definition 7.6 $\mathcal{M}_A^*$ *is called a minimal covering set of $\mathcal{M}_A$ if*

1) $\mathcal{M}_A^* \subseteq \mathcal{M}_A$;
2) $\forall M \in \mathcal{M}_A$, $\exists M' \in \mathcal{M}_A^*$ *such that $M' \succcurlyeq M$; and*
3) $\forall M \in \mathcal{M}_A^*$, $\nexists M'' \in \mathcal{M}_A^*$ *such that $M \neq M''$ and $M'' \succcurlyeq M$.*

Based on these definitions, if the markings in $\mathcal{M}_{FIM}^*$ are forbidden by the place invariants designed for a supervisor, all markings in $\mathcal{M}_{FIM}$ are also forbidden by the supervisor. Similarly, if the markings in $\mathcal{M}_A^*$ are permitted by some place designed for a supervisor, all markings in $\mathcal{M}_A$ are also permitted by the supervisor. Therefore, to design a supervisor that survives all admissible markings and forbids all FIMs, we only need to consider two reduced sets of markings: $\mathcal{M}_{FIM}^*$ and $\mathcal{M}_A^*$. For a marking $M' \in \mathcal{M}_{FIM}^* \subseteq \mathcal{M}_{FIM}$, the constraint that forbids M' and survives all admissible markings is thus simplified to:

$$\sum_{i \in \mathbb{N}_A} l(i) \cdot (M(p_i) - M'(p_i)) \leq -1, \quad \forall M \in \mathcal{M}_A^*. \tag{7.5}$$

7.2.4   Admissible and Structure-Minimal Supervisor Synthesis

This section presents an IPL method to minimize the structure of the aforementioned admissible supervisor in the presence of uncontrollable and unobservable transitions.

To minimize the structure of a supervisor, the first objective is to minimize the number of monitors in the supervisor. Note that each designed monitor corresponds to a place invariant that permits all markings in $\mathcal{M}_A^*$ and forbids at least one marking in $\mathcal{M}_{\mathrm{FIM}}^*$. In addition, the place invariants of these monitors should forbid all markings in $\mathcal{M}_{\mathrm{FIM}}^*$. Let I_j with $j \in \mathbb{N}_{\mathrm{FIM}}^* = \{i | M_i \in \mathcal{M}_{\mathrm{FIM}}^*\}$ be the place invariant that forbids the jth marking in $\mathcal{M}_{\mathrm{FIM}}^*$ and use $q_j \in \{0,1\}$ to indicate whether or not I_j is selected to construct a monitor. $q_j = 1$ represents the selection of I_j and $q_j = 0$ represents no selection. Then, the following objective can be used to obtain the fewest monitors in the supervisor synthesis:

$$\min O_1 = \sum_{j \in \mathbb{N}_{\mathrm{FIM}}^*} q_j. \tag{7.6}$$

The second objective is to minimize the number of arcs of the supervisor. We use $u_{j,y} \in \{0,1\}$ to denote whether or not an arc is added to connect a control place p_{c_j} in the supervisor to a transition $t_y \in T$. Let $[N_c]$ be the incidence matrix between control places and transitions in T and $[N_p]$ be the incidence matrix of the original net. According to Section 3.3.2, we have $[N_c] = -[L] \cdot [N_p]$ where $[L]$ is an $m_c \times |P|$ nonnegative integer matrix that denotes the coefficients of the designed place invariants. Thus, $[N_c](j,y) = -\sum_{i \in \mathbb{N}_A} l_j(i) \cdot [N_p](i,y)$. Therefore, the constraint between a place invariants and an arc is:

$$-\sum_{i \in \mathbb{N}_A} l_j(i) \cdot [N_p](i,y) \geq -\Gamma \cdot u_{j,y} - \Gamma \cdot (1 - q_j),$$

$$\forall j \in \mathbb{N}_{\mathrm{FIM}}^*, \forall y \in \mathbb{N}_T \tag{7.7}$$

where $\mathbb{N}_T = \{i | t_i \in T\}$. This constraint indicates that if $q_j = 1$ and $[N_c](j,y) \leq -1$, then $u_{j,y} = 1$, which implies an arc from p_{c_j} to t_y. Similarly, another set of variables $v_{j,y} \in \{0,1\}$ is used to denote whether or not an arc exists from t_y to p_{c_j} and we have

$$\sum_{i \in \mathbb{N}_A} l_j(i) \cdot [N_p](i,y) \geq -\Gamma \cdot v_{j,y} - \Gamma \cdot (1 - q_j),$$

$$\forall j \in \mathbb{N}_{\mathrm{FIM}}^*, \forall y \in \mathbb{N}_T. \tag{7.8}$$

This constraint indicates that if $q_j = 1$ and $[N_c](j,y) \geq 1$, then $v_{j,y} = 1$, which implies an arc from t_y to p_{c_j}. Therefore, the following objective can be used to ensure the fewest arcs in the supervisor.

$$\min O_2 = \sum_{j \in \mathbb{N}_{\mathrm{FIM}}^*} \sum_{y \in \mathbb{N}_T} (u_{j,y} + v_{j,y}). \tag{7.9}$$

In addition, no arc should exist between the supervisor and unobservable transitions; otherwise, the firings of these unobservable transitions change the marking of the supervisor, which means that the unobservable event in the system can be observed or detected by the devices that implement the supervisor in some form or another. Thus, we have

$$u_{j,y} + v_{j,y} = 0, \quad \forall j \in \mathbb{N}^*_{\text{FIM}}, \quad \forall y \in \mathbb{N}_{T_{\hat{O}}} \tag{7.10}$$

where $\mathbb{N}_{T_{\hat{O}}} = \{i | T_i \in T_{\hat{O}}\}$. As a result, an admissible and structure-minimal supervisor for a DP-net with uncontrollable and unobservable transitions can be obtained by solving the following IPL, which is denoted as a Supervisor Synthesis with Uncontrollable and Unobservable Transitions (S^2U^2T) problem:

S^2U^2T :

$$lex \min\{O_1, O_2\} \tag{7.11}$$

subject to

$$\sum_{i \in \mathbb{N}_A} l_j(i) \cdot \left(M_l(p_i) - M_j(p_i)\right) \leq -1,$$

$$\forall M_j \in \mathcal{M}^*_{\text{FIM}}, \quad \forall M_l \in \mathcal{M}^*_A \tag{7.12}$$

$$-\sum_{i \in \mathbb{N}_A} l_j(i) \cdot [N_p](i, y) \geq -\Gamma \cdot u_{j,y} - \Gamma \cdot (1 - q_j),$$

$$\forall j \in \mathbb{N}^*_{\text{FIM}}, \quad \forall y \in \mathbb{N}_T \tag{7.13}$$

$$\sum_{i \in \mathbb{N}_A} l_j(i) \cdot [N_p](i, y) \geq -\Gamma \cdot v_{j,y} - \Gamma \cdot (1 - q_j),$$

$$\forall j \in \mathbb{N}^*_{\text{FIM}}, \quad \forall y \in \mathbb{N}_T \tag{7.14}$$

$$u_{j,y} + v_{j,y} = 0, \quad \forall j \in \mathbb{N}^*_{\text{FIM}}, \quad \forall y \in \mathbb{N}_{T_{\hat{O}}} \tag{7.15}$$

$$\sum_{i \in \mathbb{N}_A} l_j(i) \cdot \left(M_k(p_i) - M_j(p_i)\right) \geq -\Gamma \cdot (1 - f_{j,k}),$$

$$\forall M_j, M_k \in \mathcal{M}^*_{\text{FIM}}, \quad j \neq k \tag{7.16}$$

$$f_{j,k} \leq q_j, \quad \forall j, k \in \mathbb{N}^*_{\text{FIM}}, \quad j \neq k \tag{7.17}$$

$$q_j + \sum_{k \in \mathbb{N}^*_{\text{FIM}}, k \neq j} f_{k,j} \geq 1, \quad \forall j \in \mathbb{N}^*_{\text{FIM}} \tag{7.18}$$

$$l_j(i) \in \{0, 1, 2, \dots\}, \quad \forall i \in \mathbb{N}_A, \quad \forall j \in \mathbb{N}^*_{\text{FIM}}$$

$$f_{j,k} \in \{0, 1\}, \quad \forall j, k \in \mathbb{N}^*_{\text{FIM}}, \quad j \neq k$$

$$q_j \in \{0, 1\}, \quad \forall j \in \mathbb{N}^*_{\text{FIM}}$$

$$u_{j,y}, v_{j,y} \in \{0, 1\}, \quad \forall j \in \mathbb{N}^*_{\text{FIM}}, \quad \forall y \in \mathbb{N}_T$$

In S^2U^2T, (7.16)–(7.18) are employed to compute all needed place invariants as MMP in Section 3.3.4. Note that the coefficients $l_j(i)$ of the place invariants should be nonnegative, which means that the designed place invariants are P-semiflows.

7.3 Deadlock Prevention Policy

In this section, a deadlock prevention policy that uses S^2U^2T to obtain admissible and structure-minimal supervisors for the DP-nets with uncontrollable and unobservable transitions is presented. Also, the condition under which the obtained supervisor is optimal is given.

Algorithm 7.2
```
Deadlock prevention policy by using the S²U²T.
Input: A DP-net with T_C̃ = T_OC̃ ∪ T_Õ.
Output: A deadlock-free controlled net with an admissible and
structure-minimal supervisor if such a supervisor exists.
Deadlock_prevention_for_S²U²T {
        Generate G(N, M₀);
        Compute M_A and M_Ã by Algorithm 7.1 in Section 7.2.2;
        Compute M*_A and M*_FIM as Section 7.2.3;
        V_P := ∅, V_A := ∅;
        Solve the S²U²T in Section 7.2.4. If no solution exists,
exit;
        For each q_j = 1 in the solution
        {
                Use l_j to compute a monitor that has a control place p_cⱼ
and a set of arcs A_j as Section 3.3.2;
                V_P := V_P ∪ {p_cⱼ}, V_A := V_A ∪ A_j;
        }
        Add V_P and V_A to (N, M₀) and output the controlled net;
}
```

Theorem 7.1 *Algorithm 7.2 obtains a deadlock-free and admissible supervisor with the fewest control places and arcs for a DP-net with uncontrollable and unobservable transitions if S^2U^2T has an optimal solution.*

Proof: S^2U^2T prevents the controlled net from entering $\mathcal{M}_{\tilde{A}}$ by using Constraint (7.18), which ensures that each marking in $\mathcal{M}^*_{FIM}$ is prohibited by at least one designed place invariant. So, the controlled net does not reach deadlocks and markings doomed to deadlock. That is to say, the added supervisor is deadlock-free. On the other hand, Constraint (7.15) enforces that no arc is added between a *control place and an unobservable transition. So, unobservable events represented by unobservable transitions are not observed or detected by the added supervisor. In addition, Constraint (7.12) permits all admissible markings that do not lead the underlying system to inadmissible markings via the firings of uncontrollable transitions. Thus, the supervisor is admissible in the presence of uncontrollable and unobservable transitions. $S^2 U^2 T$ minimizes the number of control places and the number of arcs in the supervisor. Therefore, if $S^2 U^2 T$ has an optimal solution, Algorithm 7.2 can obtain a deadlock-free and admissible supervisor that has a minimal structure regarding both control places and added arcs.* ∎

In addition, the controlled net is reversible since its reachability graph is an SCC that contains M_0. Next, we give the conditions under which the supervisors obtained by the presented method are optimal (maximally permissive in behavior). To show it, the definition of crucial transitions (Chen & Barkaoui, 2014) should first be introduced.

Definition 7.7 *For a net system (N, M_0), the set of marking/transition separation instances (MTSIs) is defined as:*

$$\Phi = \{(M, t)|M[t\rangle M' \wedge M \in \mathcal{M}_L \wedge M' \in \mathcal{M}_D\}, \tag{7.19}$$

and the set of crucial transitions is defined as:

$$T_{cru} = \{t \in T|\exists M \in R(N, M_0) \text{ s.t. } (M, t) \in \Phi\}. \tag{7.20}$$

Theorem 7.2 *The supervisor obtained by Algorithm 7.2 is optimal if all crucial transitions are controllable, i.e., $T_{cru} \subseteq T_C$.*

Proof: According to Theorem 7.1, the supervisor obtained by Algorithm 7.2 forbids all FIMs and permits all admissible markings. By Algorithm 7.1, all admissible markings and the edges among them form an SCC that contains M_0 in the reachability graph $G(N, M_0)$. If $T_{cru} \subseteq T_C$, there is no uncontrollable dangerous marking in $G(N, M_0)$. According to Algorithm 7.1, the SCC includes all legal markings in the live zone and excludes all markings in the deadlock zone of $G(N, M_0)$, i.e., $\mathcal{M}_A = \mathcal{M}_L$. By Definition 3.2, the supervisor is optimal. ∎

Note that a crucial transition, which leads the system from the live zone to the deadlock zone, may be uncontrollable or unobservable. The method presented in this chapter can deal with such DP-nets. To illustrate it, we use the DP-net in Figure 7.1 as an example. In this net, $T_{O\tilde{C}} = \{t_1, t_5, t_9\}$ and $T_{\tilde{O}} = \{t_3, t_8\}$. Its reachability graph is given in Figure 7.2 in which $T_{cru} = \{t_1, t_2, t_7\}$. It cannot be handled by the method in Chen and Barkaoui (2014) since t_1 is both crucial and uncontrollable. By using the method of this chapter, the reachability graph contains 30 reachable markings, 25 of which are admissible markings and 5 of which are FIMs. In addition, we have $\mathcal{M}_A^* = \{M_5 = p_2 + p_3 + p_4, M_6 = p_2 + p_3 + p_5, M_{14} = p_2 + p_7, M_{17} = p_2 + p_4 + p_8, M_{18} = p_2 + p_5 + p_8, M_{23} = p_7 + p_9, M_{24} = p_8 + p_9, M_{28} = p_5 + p_9, M_{29} = p_4 + p_9\}$ and $\mathcal{M}_{FIM}^* = \{M_{11} = p_3 + p_7, M_{21} = p_7 + p_8\}$. The formulated S^2U^2T of the model, which has 64 constraints and 58 variables, is given as follows.

S^2U^2T :

$$lex\min\left\{\sum_{j\in\mathbb{N}_{FIM}^*} q_j, \sum_{j\in\mathbb{N}_{FIM}^*}\sum_{y\in\mathbb{N}_T}(u_{j,y} + v_{j,y})\right\}$$

subject to

$$l_{1,1} + l_{1,3} - l_{1,5} \le -1$$
$$l_{1,1} + l_{1,4} - l_{1,5} \le -1$$
$$l_{1,1} - l_{1,2} \le -1$$
$$l_{1,1} - l_{1,2} + l_{1,3} - l_{1,5} + l_{1,6} \le -1$$
$$l_{1,1} - l_{1,2} + l_{1,4} - l_{1,5} + l_{1,6} \le -1$$
$$-l_{1,2} + l_{1,7} \le -1$$
$$-l_{1,2} - l_{1,5} + l_{1,6} + l_{1,7} \le -1$$
$$-l_{1,2} + l_{1,4} - l_{1,5} + l_{1,7} \le -1$$
$$-l_{1,2} + l_{1,3} - l_{1,5} + l_{1,7} \le -1$$
$$l_{2,1} + l_{2,2} + l_{2,3} - l_{2,5} - l_{2,6} \le -1$$
$$l_{2,1} + l_{2,2} + l_{2,4} - l_{2,5} - l_{2,6} \le -1$$
$$l_{2,1} - l_{2,6} \le -1$$
$$l_{2,1} + l_{2,3} - l_{2,5} \le -1$$
$$l_{2,1} + l_{2,4} - l_{2,5} \le -1$$
$$-l_{2,6} + l_{2,7} \le -1$$
$$-l_{2,5} + l_{2,7} \le -1$$
$$l_{2,4} - l_{2,5} - l_{2,6} + l_{2,7} \le -1$$
$$l_{2,3} - l_{2,5} - l_{2,6} + l_{2,7} \le -1$$
$$-l_{1,2} + l_{1,6} \ge -\Gamma \cdot (1 - f_{1,2})$$
$$l_{2,2} - l_{2,6} \ge -\Gamma \cdot (1 - f_{2,1})$$

$$f_{1,2} \leq q_1$$

$$f_{2,1} \leq q_2$$

$$f_{2,1} + q_1 \geq 1$$

$$f_{1,2} + q_2 \geq 1$$

$$-l_{1,1} \geq -\Gamma \cdot (u_{1,1} + 1 - q_1)$$

$$l_{1,1} - l_{1,2} \geq -\Gamma \cdot (u_{1,2} + 1 - q_1)$$

$$l_{1,2} - l_{1,4} \geq -\Gamma \cdot (u_{1,4} + 1 - q_1)$$

$$l_{1,3} - l_{1,4} \geq -\Gamma \cdot (u_{1,5} + 1 - q_1)$$

$$l_{1,4} \geq -\Gamma \cdot (u_{1,6} + 1 - q_1)$$

$$-l_{1,5} \geq -\Gamma \cdot (u_{1,7} + 1 - q_1)$$

$$l_{1,6} - l_{1,7} \geq -\Gamma \cdot (u_{1,9} + 1 - q_1)$$

$$l_{1,7} \geq -\Gamma \cdot (u_{1,10} + 1 - q_1)$$

$$-l_{2,1} - \Gamma \cdot q_2 \geq -\Gamma \cdot (u_{2,1} + 1 - q_2)$$

$$l_{2,1} - l_{2,2} \geq -\Gamma \cdot (u_{2,2} + 1 - q_2)$$

$$l_{2,2} - l_{2,4} \geq -\Gamma \cdot (u_{2,4} + 1 - q_2)$$

$$l_{2,3} - l_{2,4} \geq -\Gamma \cdot (u_{2,5} + 1 - q_2)$$

$$l_{2,4} \geq -\Gamma \cdot (u_{2,6} + 1 - q_2)$$

$$-l_{2,5} \geq -\Gamma \cdot (u_{2,7} + 1 - q_2)$$

$$l_{2,6} - l_{2,7} \geq -\Gamma \cdot (u_{2,9} + 1 - q_2)$$

$$l_{2,7} \geq -\Gamma \cdot (u_{2,10} + 1 - q_2)$$

$$l_{1,1} - \Gamma \cdot q_1 \geq -\Gamma \cdot (v_{1,1} + 1 - q_1)$$

$$-l_{1,1} + l_{1,2} \geq -\Gamma \cdot (v_{1,2} + 1 - q_1)$$

$$-l_{1,2} + l_{1,4} \geq -\Gamma \cdot (v_{1,4} + 1 - q_1)$$

$$-l_{1,3} + l_{1,4} \geq -\Gamma \cdot (v_{1,5} + 1 - q_1)$$

$$-l_{1,4} \geq -\Gamma \cdot (v_{1,6} + 1 - q_1)$$

$$l_{1,5} \geq -\Gamma \cdot (v_{1,7} + 1 - q_1)$$

$$-l_{1,6} + l_{1,7} \geq -\Gamma \cdot (v_{1,9} + 1 - q_1)$$

$$-l_{1,7} \geq -\Gamma \cdot (v_{1,10} + 1 - q_1)$$

$$l_{2,1} \geq -\Gamma \cdot (v_{2,1} + 1 - q_2)$$

$$-l_{2,1} + l_{2,2} \geq -\Gamma \cdot (v_{2,2} + 1 - q_2)$$

$$-l_{2,2} + l_{2,4} \geq -\Gamma \cdot (v_{2,4} + 1 - q_2)$$

$$-l_{2,3} + l_{2,4} \geq -\Gamma \cdot (v_{2,5} + 1 - q_2)$$

$$-l_{2,4} \geq -\Gamma \cdot (v_{2,6} + 1 - q_2)$$

$$l_{2,5} \geq -\Gamma \cdot (v_{2,7} + 1 - q_2)$$

$$-l_{2,6} + l_{2,7} \geq -\Gamma \cdot (v_{2,9} + 1 - q_2)$$

$$-l_{2,7} \geq -\Gamma \cdot (v_{2,10} + 1 - q_2)$$

$$l_{1,1} - l_{1,3} \geq -\Gamma \cdot (1 - q_1)$$

$$l_{1,5} - l_{1,6} \geq -\Gamma \cdot (1 - q_1)$$

$$l_{2,1} - l_{2,3} \geq -\Gamma \cdot (1 - q_2)$$

$$l_{2,5} - l_{2,6} \geq -\Gamma \cdot (1 - q_2)$$

$$-l_{1,1} + l_{1,3} \geq -\Gamma \cdot (1 - q_1)$$

$$-l_{1,5} + l_{1,6} \geq -\Gamma \cdot (1 - q_1)$$

$$-l_{2,1} + l_{2,3} \geq -\Gamma \cdot (1 - q_2)$$

$$-l_{2,5} + l_{2,6} \geq -\Gamma \cdot (1 - q_2)$$

$$l_j(i) \in \{0, 1, 2, \dots\}, \quad \forall i \in \mathbb{N}_A, \quad \forall j \in \mathbb{N}_{\text{FIM}}^*$$

$$f_{j,k} \in \{0, 1\}, \quad \forall j, k \in \mathbb{N}_{\text{FIM}}^*, \quad j \neq k$$

$$q_j \in \{0, 1\}, \quad \forall j \in \mathbb{N}_{\text{FIM}}^*$$

$$u_{j,y}, v_{j,y} \in \{0, 1\}, \quad \forall j \in \mathbb{N}_{\text{FIM}}^*, \quad \forall y \in \mathbb{N}_T$$

$$\mathbb{N}_A = \{1, 2, 3, \dots, 7\}$$

$$\mathbb{N}_{\text{FIM}}^* = \{1, 2\}$$

$$\mathbb{N}_T = \{1, 2, 3, \dots, 10\}$$

In S^2U^2T, $l_{j,i}$ is used to denote $l_j(i)$ for expedience. S^2U^2T has a solution with $q_1 = f_{1,2} = l_{1,2} = l_{1,5} = l_{1,6} = u_{1,2} = u_{1,7} = v_{1,4} = v_{1,9} = 1$ and the rest variables equaling zero. Since $q_1 = 1$ and $q_2 = 0$, only the place invariant that forbids the first marking in $\mathcal{M}_{\text{FIM}}^*$ is selected to design a control place p_{c_1} and its connected arcs to form a supervisor. By using the supervisor computation method in Section 3.3.2, we have $^\bullet p_{c_1} = \{t_4, t_9\}$, $p_{c_1}^\bullet = \{t_2, t_7\}$, and $M_0(p_{c_1}) = 1$. After adding the supervisor to the original net, we get a controlled net as shown in Figure 7.3. The added supervisor is admissible and deadlock-free since all admissible markings are reachable in the controlled net and all five inadmissible markings, including two deadlocks (M_9 and M_{22}), a marking doomed to deadlock (M_{11}), an uncontrollable dangerous marking (M_{21}), and a marking (M_{25}) that inevitably evolves into an uncontrollable dangerous marking, are forbidden by the supervisor. In addition, it guarantees that the supervisor has the fewest control places and arcs, i.e., its structure is minimal.

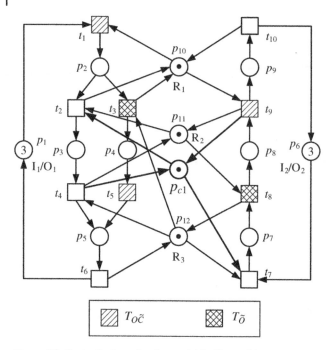

Figure 7.3 Controlled net for the model in Figure 7.1.

7.4 Illustrative Experiments

This section tests some RAS benchmarks from the literature in a computer with Intel i3 Core 2.93 GHz CPU and 4GB memory to demonstrate the presented method. We use INA to generate $G(N, M_0)$ and develop some C++ programs to compute $\mathcal{M}_A$, $\mathcal{M}_{\text{FIM}}$, $\mathcal{M}_A^*$, and $\mathcal{M}_{\text{FIM}}^*$. Then, Lingo is adopted to solve S^2U^2T. Finally, INA is used again to verify the behavioral properties of the controlled nets.

First, we consider a flexible manufacturing system in Ghaffari *et al.* (2003) and Chen and Li (2011), which has two robots R_1 and R_2, three kinds of machines M_1-M_3, two loading buffers I_1 and I_2, and two unloading buffers O_1 and O_2. Three kinds of parts P_1-P_3 are handled with the following sequences:

$$P_1: I_1 \rightarrow R_1 \rightarrow M_3 \rightarrow O_1$$
$$P_2: I_1 \rightarrow R_1 \rightarrow M_1 \rightarrow R_2 \rightarrow M_2 \rightarrow O_1$$
$$P_3: I_2 \rightarrow M_2 \rightarrow R_2 \rightarrow M_1 \rightarrow O_2$$

Figure 7.4 shows its DP-net that has 11 transitions and 15 places in which $P_0 = \{p_1, p_{10}\}$, $P_R = \{p_{11} - p_{15}\}$, and $P_A = \{p_2 - p_9\}$. Suppose that $T_{O\tilde{C}} = \{t_0, t_3, t_4, t_5, t_7, t_9\}$ and $T_{\tilde{O}} = \{t_6, t_8\}$ that are depicted in Figure 7.4 by

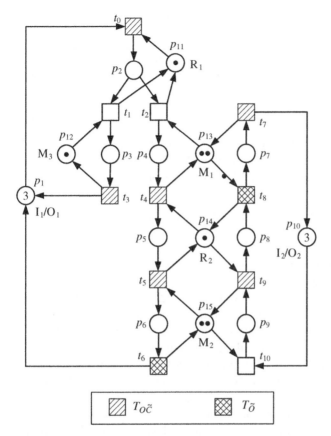

Figure 7.4 DP-net of a flexible manufacturing system (Chen & Li, 2011; Ghaffari *et al.*, 2003).

different squares. Its reachability graph has 261 reachable markings (232 of which are in the live zone), 215 admissible markings (22 of which belong to $\mathcal{M}_A^*$), and 37 FIMs (four of which are in $\mathcal{M}_{FIM}^*$). For this net, the formulated S^2U^2T has 204 constraints and 136 variables. It is solved within eight seconds and the obtained supervisor has two control places and eight arcs whose details are shown in Table 7.1. After adding the supervisor to the original net, the obtained controlled net is deadlock-free and reversible. In addition, it is admissible since all 215 admissible markings are reachable in the controlled net in the presence of uncontrollable and unobservable transitions. The obtained supervisor has one control place and four arcs fewer than the supervisor obtained by the method in Ghaffari *et al.* (2003). Most importantly, the supervisor is guaranteed to be structurally minimal in terms of both control places and added arcs.

Table 7.1 Supervisor for the net in Figure 7.4 with $T_{O\check{C}} = \{t_0, t_3, t_4, t_5, t_7, t_9\}$ and $T_{\check{O}} = \{t_6, t_8\}$.

j	FIM_j	I_j	${}^{\bullet}p_{c_j}$	$p_{c_j}^{\bullet}$	$M_0(p_{c_j})$
1	$2p_4 + p_9$	$3M_4 + M_7 + M_8 + M_9 + M_{p_{c_1}} = 6$	$3t_4, t_7$	$3t_2, t_{10}$	6
4	$p_4 + 2p_9$	$2M_4 + 2M_5 + 4M_9 + M_{p_{c_4}} = 9$	$2t_5, 4t_9$	$2t_2, 4t_{10}$	9

Then, we consider another manufacturing system (Li *et al.*, 2008a; Uzam, 2002) that has two robots R_1 and R_2, four machines M_1–M_4, two loading buffers I_1 and I_2, and two unloading buffers O_1 and O_2. Two kinds of parts are to be processed with the following production sequences:

$$P_1: I_1 \rightarrow R_1 \rightarrow M_1(\text{or } M_2) \rightarrow R_1 \rightarrow M_3 \rightarrow R_2 \rightarrow O_1$$
$$P_2: I_2 \rightarrow R_2 \rightarrow M_4 \rightarrow R_1 \rightarrow M_2 \rightarrow R_1 \rightarrow O_2$$

Figure 7.5 shows its DP-net that has 14 transitions and 19 places with $P_0 = \{p_1, p_8\}$, $P_R = \{p_{14} - p_{19}\}$, and $P_A = \{p_2 - p_7, p_9 - p_{13}\}$. Suppose that $T_{O\check{C}} = \{t_3, t_7, t_8, t_{10}, t_{13}, t_{14}\}$ and $T_{\check{O}} = \{t_6, t_{12}\}$ that are depicted in Figure 7.5 by different squares. We have $T_{\check{C}} = \{t_3, t_6, t_7, t_8, t_{10}, t_{12}, t_{13}, t_{14}\}$ and $T_C = \{t_1, t_2, t_4, t_5, t_9, t_{11}\}$. Its reachability graph has 282 reachable markings in which $|\mathcal{M}_L| = 205$ and $|\mathcal{M}_{FIM}| = 54$. Note that, for this net, all markings in the live zone are admissible, i.e., $\mathcal{M}_A = \mathcal{M}_L$. $\mathcal{M}_A^*$ and $\mathcal{M}_{FIM}^*$ have 26 and 8 markings, respectively. Its formulated S^2U^2T has 552 constraints and 376 variables. The solution time is 62 seconds and the obtained supervisor has two control places and 12 added arcs as shown in Table 7.2. The supervisor is deadlock-free, admissible, and structurally minimal, and it controls no uncontrollable transitions and observes no unobservable transitions. Furthermore, (M_{43}, t_9), (M_{51}, t_9), (M_{55}, t_9), (M_{56}, t_9), (M_{59}, t_9), (M_{60}, t_9), (M_{61}, t_9), (M_{145}, t_9), and (M_{171}, t_9) are MTSIs. Thus, we have $T_{cru} = \{t_9\}$ and $T_{cru} \subseteq T_C$. Thus, the obtained supervisor is also optimal in behavior according to Theorem 7.2. Note that the method in Chen and Barkaoui (2014) can also handle this net. However, its obtained supervisor is not minimal in structure and it cannot deal with the nets whose crucial transitions may be uncontrollable or unobservable.

To illustrate this, we consider the net in Figure 7.5 with $T_{O\check{C}} = \{t_3, t_7, t_9, t_{10}\}$ and $T_{\check{O}} = \{t_6, t_{13}\}$. We have $T_{\check{C}} = \{t_3, t_6, t_7, t_9, t_{10}, t_{13}\}$ and the rest transitions belong to T_C. Its reachability graph has 282 reachable markings, 205 of which are in the live zone, 160 of which are admissible markings, and 66 of which are FIMs. For this net, $T_{cru} = \{t_9\}$ and $T_{cru} \nsubseteq T_C$. So, it cannot be handled by the method in Chen and Barkaoui (2014) since t_9 is both crucial and uncontrollable. By using the method presented in this chapter, $\mathcal{M}_A^*$ and $\mathcal{M}_{FIM}^*$ have 19 and 8 markings, respectively. Then, an S^2U^2T that consists of 496 constraints and 376 variables is formulated. The solution time of the S^2U^2T is about 14 seconds and the obtained supervisor has

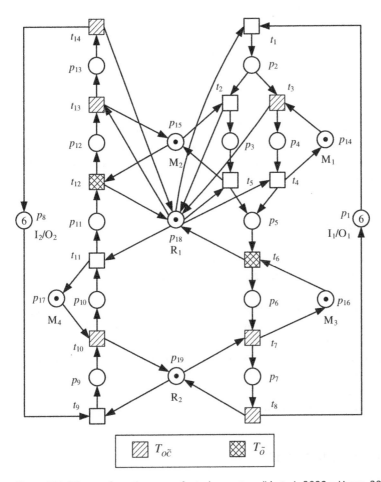

Figure 7.5 DP-net of another manufacturing system (Li *et al.*, 2008a; Uzam, 2002).

Table 7.2 Supervisor for the net in Figure 7.5 with $T_{o\bar{c}} = \{t_3, t_7, t_8, t_{10}, t_{13}, t_{14}\}$ and $T_{\bar{o}} = \{t_6, t_{12}\}$.

j	FIM_j	l_j	$\bullet p_{c_j}$	$p_{c_j}^\bullet$	$M_0(p_{c_j})$
2	$p_3 + p_5 + p_9 + p_{10}$	$4M_3 + 4M_5 + 9M_9 + 9M_{10} +$ $M_{p_{c_2}} = 25$	$4t_8, 9t_{11}$	$3t_1, t_2, t_4, 9t_9$	25
8	$p_3 + p_{11}$	$2M_3 + 2M_{11} + M_{p_{c_8}} = 3$	$t_4, 2t_5, 2t_{13}$	$t_1, t_2, 2t_{11}$	3

Table 7.3 Supervisor for the net in Figure 7.5 with $T_{0\dot{c}} = \{t_3, t_7, t_9, t_{10}\}$ and $T_{\dot{0}} = \{t_6, t_{13}\}$.

j	FIM_j	l_j	$^\bullet p_{c_j}$	$p_{c_j}^\bullet$	$M_0(p_{c_j})$
2	$p_3 + p_5$	$2M_3 + 2M_5 + M_{p_{c_2}} = 3$	$2t_7$	t_1, t_2, t_4	3
5	$p_{11} + p_{12}$	$2M_{11} + 2M_{12} + M_{p_{c_5}} = 3$	$t_4, 2t_5, 2t_{14}$	$t_1, t_2, 2t_{11}$	3

2 places and 10 arcs as shown in Table 7.3. The controlled net is admissible since it survives all 160 admissible markings of the original net. It is also deadlock-free and reversible. In addition, the added supervisor is ensured to be structurally minimal in terms of both places and arcs.

7.5 Concluding Remarks

This chapter presents an ILP-based deadlock prevention method to design admissible and structure-minimal supervisors for DP-nets with uncontrollable and unobservable transitions. It can deal with the nets whose crucial transitions may be uncontrollable and/or unobservable. Furthermore, its obtained supervisors are guaranteed to be deadlock-free, admissible, and structurally minimal in terms of both places and arcs. The condition under which the obtained supervisor is behaviorally maximal is also given. In the future, how to develop speed-up techniques for the presented method, such as symbolic representations and GPU computing methods, to efficiently generate reachable markings and other net structures is an interesting issue to address.

7.6 Bibliographical Notes

In practice, controllability and observability are two important event properties in RASs. This chapter presents a deadlock prevention method based on Petri nets with uncontrollable and unobservable events for RASs. The well-known supervisor computation method based on place invariants is proposed in Yamalidou *et al.* (1996). Seminal work on the theory of regions that designs optimal and near-optimal Petri net supervisors can be found in Ghaffari *et al.* (2003), Uzam (2002) and Uzam and Zhou (2006, 2007). An optimal supervisor synthesis method by solving an ILP can be seen in Chen and Li (2011). For the classifications of transitions in Petri nets with uncontrollability and unobservability, please refer to Giua *et al.* (1992), Moody and Antsaklis (2000), and Qin *et al.* (2012). The concept of crucial transitions is pioneered in Chen and Barkaoui (2014). Most material reported in this chapter on designing admissible and minimal supervisors with uncontrollability and unobservability comes from Huang *et al.* (2017).

Part III

Heuristic Scheduling

8

Informed Heuristic Search in Reachability Graph

The scheduling of a system is a decision-making process that plays an important role in improving the performance of resource allocation systems (RASs). Timed Petri nets are suitable for graphically and concisely modeling such systems with time information and obtaining their state-space structures. Within their reachability graphs, Petri nets' evolution and intelligent search algorithms can be combined to find an efficient transition firing sequence from an initial state to a goal state of the underlying system. First, this chapter presents two methods to model RASs with place-timed Petri nets, i.e., converting existing Petri net models of RASs in the literature and synthesizing new models via a top-down or bottom-up method. Then, the state evolution of such nets is introduced. Finally, an algorithm that combines Petri nets' evolution, a heuristic A* search, and state check rules is presented to schedule the nets of RASs without exploring their whole reachability graphs.

8.1 Introduction

Resource allocation systems (RASs), such as automated manufacturing systems, project management systems, and multithreaded software engineering systems, are a class of concurrent systems in which some shared resources are granted to a set of concurrent processes. RASs often provide a great number of choices of resources and alternative plans, which allow high system efficiency. The scheduling of RASs is to select a processing sequence among alternative plans and assign resources to activities to achieve the best efficiency in the systems. However, the scheduling problem is known to be complex even for simple RASs and is NP-hard in many cases (Tuncel & Bayhan, 2007). In addition, RASs with routing flexibility and limited buffer capacity may bring the systems to deadlocks (Mejía *et al.*, 2018).

Supervisory Control and Scheduling of Resource Allocation Systems: Reachability Graph Perspective, First Edition. Bo Huang and MengChu Zhou.

A complete and general tool for the RAS scheduling problem should have the ability to well model the underlying systems and provide an efficient and general technique to schedule them. Petri nets (PNs) are very suitable for it (Ding *et al.*, 2013; Du *et al.*, 2014; Tuncel & Bayhan, 2007; Zhou & DiCesare, 1993). Petri nets can graphically and concisely model concurrent and asynchronous activities, multi-resource sharing, routing flexibility, limited buffers, and complex constraints which are common in RASs and have an underlying mathematical foundation to perform qualitative and quantitative analysis of the models.

In addition, Petri nets with time information, called timed Petri nets, can generate a state-space structure, with which the scheduling problem can be solved by combining timed Petri nets' evolution with dispatching rules or intelligent search algorithms. Among them, intelligent search algorithms have been paid more attention than the dispatching rules such as shortest total processing time, first-come-first-served, and shortest imminent processing time (Li *et al.*, 2013). For example, in Lee and DiCesare (1994), a novel method that combines a heuristic A* algorithm with the timed Petri net execution is proposed to schedule flexible manufacturing systems. This method only needs to explore a partial reachability graph to obtain a good transition firing sequence from an initial marking to a goal marking, i.e., a good executable schedule of the underlying system. In addition, if the heuristic function used by the A* algorithm is admissible, the obtained schedule is guaranteed to be optimal. This method is a basis for many related studies (Baruwa *et al.*, 2015; Huang *et al.*, 2008, 2014; Lee & Lee, 2010; Luo *et al.*, 2015; Mejía & Odrey, 2005; Mejía *et al.*, 2018; Moro *et al.*, 2002b; Wang *et al.*, 2019; Xiong & Zhou, 1998). This chapter introduces the procedures on constructing place-timed Petri nets for RASs, the state evolution of the place-timed Petri nets, and the A* search within the reachability graphs to schedule RASs.

8.2 System Scheduling with Place-Timed Petri Nets

8.2.1 Place-Timed Petri Nets

Petri nets are widely used to model and analyze RASs (Chen *et al.*, 2017; Hrúz & Zhou, 2007; Liu *et al.*, 2010; Moody & Antsaklis, 2012; Seatzu *et al.*, 2012; Wu & Zhou, 2007; Wu *et al.*, 2008, 2009; Zhou & DiCesare, 1993; Zhou & Jeng, 1998; Zhu *et al.*, 2018). However, Petri nets have no time information and their definitions are thus not enough for the scheduling of RASs. The scheduling of RASs necessitates the introduction of Petri nets with time information, i.e., timed Petri nets (Hu *et al.*, 2010; Qiao, *et al.*, 2013) or time Petri nets (Li *et al.*, 2020).

Timed Petri nets can be divided into two classes: deterministic timed Petri nets whose components are associated with deterministic firing times or time intervals and stochastic timed Petri nets whose components are associated with random firing times or time intervals (Zhou *et al.*, 2019). Both of them can be further divided into several subclasses.

The subclasses of deterministic timed Petri nets include place-timed Petri nets whose places are associated with deterministic operation times, transition-timed Petri nets whose transitions are associated with deterministic firing times or firing time intervals, and arc-timed Petri nets whose directed arcs are associated with deterministic times. In practice, deterministic timed Petri nets can be used to obtain production time, identify bottleneck resources, verify timing constraints, etc.

The subclasses of stochastic timed Petri nets include stochastic Petri nets whose transitions are associated with exponentially distributed firing times, generalized stochastic Petri nets whose transitions are associated with exponentially distributed firing times and firing delays that equal zero, extended stochastic Petri nets whose transitions are associated with exponentially distributed or arbitrarily distributed firing times, deterministic–stochastic Petri nets whose transitions are associated with exponentially distributed or deterministic firing times, and arbitrary stochastic Petri nets whose firing times of transitions are arbitrarily distributed. Stochastic timed Petri nets are usually used to derive production rate, throughput, average delays, critical resource utilization, etc.

For both classes of timed Petri nets, there exist subclasses of high-level Petri net with compact structures, such as timed colored Petri nets, colored stochastic Petri nets, and stochastic high-level Petri nets.

Place-timed Petri nets are often used to model and schedule RASs that are a typical kind of discrete event systems (Baruwa *et al.*, 2015; Huang *et al.*, 2008, 2014; Lee & DiCesare, 1994; Luo *et al.*, 2015; Mejía & Odrey, 2005; Mejía *et al.*, 2018; Moro *et al.*, 2002b; Xiong & Zhou, 1998). In the sequel, we use them to model and schedule RASs. The definition of place-timed Petri nets can be given as follows:

Definition 8.1 *A place-timed Petri net is a five-tuple $\mathcal{N} = (P, T, F, W, D)$ in which $P = \{p_1, p_2, ..., p_m\}$ and $T = \{t_1, t_2, ..., t_n\}$ are the finite non-empty sets of places and transitions such that $m, n \in \mathbb{Z}^+$, $P \cap T = \emptyset$, $F \subseteq (P \times T) \cup (T \times P)$ is a set of directed arcs connecting places and transitions, $W : F \to \mathbb{Z}^+$ is a weight assignment for all arcs, and D is a set of time delays associated with places.*

For RASs, two methods exist to obtain their place-timed Petri net models, i.e., converting from an existing untimed Petri net model and synthesizing from scratch. These two methods are introduced in the next two sections, respectively.

8.2.2 Conversion from an Untimed Petri Net

To model RASs, many untimed Petri nets have been used in the literature, for example, PPN, S^3PR, ES^3PR, LS^3PR, ELS^3PR, GLS^3PR, WS^3PR, S^3PGR^2, WS^3PSR, S^4R, S^4PR, G-system, S^*PR, S^5PR, and PC^2R. For details of these nets, please refer to Section 2.2.5. Such nets can model different kinds of RASs in which some resources need to be allocated to different operations. The main differences among them are the structures of the processing subnets and the number and types of resources that can be allocated to the operations. However, none of them have time information that is important to the scheduling of RASs. To model and schedule RASs, we use place-timed Petri nets whose times are associated with places to indicate the time that must elapse between the arrival of a token in a place and when it is permitted to participate in firing a post-transition of the place. The aforementioned untimed Petri nets in literature can be converted into place-timed nets after dividing each idle place (if any) into a start place and an end place in its processing subnet. Then, tokens in the idle place are put into the start place and time delays are added onto activity places. The resultant place-timed Petri nets are defined in Definition 8.2, denoted as system scheduling nets (SC-nets) for RASs.

Definition 8.2 *An SC-net is a place-timed Petri net* $\mathcal{N} = (P_S \cup P_E \cup P_R \cup P_A, T, F, W, D)$, *which is defined as the composition of a set of processing subnets* $\mathcal{N}^x = (P_S^x \cup P_E^x \cup P_R^x \cup P_A^x, T^x, F^x, W^x, D^x)$ *sharing some places, i.e.,* $\mathcal{N} = \bigcup_{x \in \mathbb{N}_J} \mathcal{N}^x$ *where* $\mathbb{N}_J = \{ i \in \mathbb{Z}^+ | \mathcal{N}^i$ *is the i-th processing subnet of the system} and the following statements are satisfied:*

1) P_S^x *(respectively, P_E^x) is called the set of start (respectively, end) places of $\mathcal{N}^x$. P_A^x (respectively, P_R^x) is called the set of activity (respectively, resource) places of $\mathcal{N}^x$.*
2) $P_A^x \neq \emptyset; P_R^x \neq \emptyset; P_S^x \not\subseteq P_A^x; P_E^x \not\subseteq P_A^x;$ *and* $(P_S^x \cup P_E^x \cup P_A^x) \cap P_R^x = \emptyset.$
3) $W = W_A \cup W_R$ *with* $W_A : F \cap ((P_A \cup P_S \cup P_E) \times T) \cup (T \times (P_A \cup P_S \cup P_E)) \rightarrow \{1\}$ *and* $W_R: F \cap ((P_R \times T) \cup (T \times P_R)) \rightarrow \mathbb{Z}^+.$
4) $\forall r \in P_R$, *there exists a unique minimal P-semiflow I_r such that* $\|I_r\| \cap P_R = \{r\}$, $\|I_r\| \cap (P_A \cup P_S \cup P_E) \neq \emptyset$, *and* $I_r(r) = 1$. *In addition,* $P_A \subseteq \bigcup_{r \in P_R} (\|I_r\| \setminus \{r\}).$
5) *The places shared by $\mathcal{N}^x$ and $\mathcal{N}^y (x \neq y)$ belong to P_R.*
6) $D: P_A \rightarrow \mathbb{N}$ *is an assignment of operation times to activity places.*

In Definition 8.2, Item 3 denotes that the weights of the arcs connecting transitions with activity places, start places, and end places are equal to one and the weights of the arcs connecting transitions with resource places are greater than one. In addition, no transition of a subnet connects an activity, start, or end place in other subnets via any arc. Item 4 represents resource preservation of the system. SC-nets are defined for the modeling and scheduling of RASs. When compared with DP-nets as defined in Section 3.2 for the deadlock handling for RASs, SC-nets can deal with PC^2R nets for software engineering. Example 8.1 shows the conversion of an S^3PR net in the literature into an SC-net.

Example 8.1 Consider an S^3PR net of a flexible manufacturing system in Huang *et al.* (2015c, 2018a), Li *et al.* (2008a), Piroddi *et al.* (2008), and Uzam (2002), which

(a)

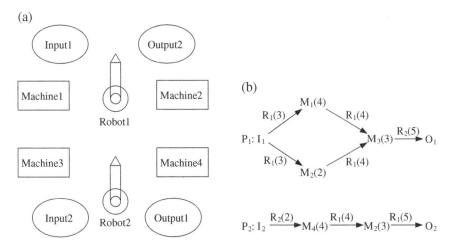

Figure 8.1 (a) A flexible manufacturing system; (b) its process sequences.

is composed of two robots R_1 and R_2 each of which holds one part at a time, four machines M_1–M_4, each of which processes one part at a time, two loading buffers I_1 and I_2, and two unloading buffers O_1 and O_2. The action areas of robot R_1 are I_1, M_1–M_4, and O_2, and the areas of robot R_2 are I_2, M_3, M_4, and O_1. As shown in Figure 8.1, two part types, P_1 and P_2, are processed in the system. P_1 is taken from I_1 by R_1, and after being handled by M_1, M_2, and M_3, it is moved to O_1 by R_2. P_2 is taken from I_2 by R_2, and after being handled by M_4 and M_2, it is moved to O_2 by R_1. Assume that there are six parts of each part type to be processed.

Its untimed S^3PR net is shown in the left of Figure 8.2 where idle places are $\{p_1, p_8\}$, resource places $P_R = \{p_{14} - p_{19}\}$, and activity (operation) places $P_A = \{p_2 - p_7, p_9 - p_{13}\}$. Suppose that the processing time required by each operation is given in parenthesis in Figure 8.1b. The net conversion and the resulting SC-net are shown in the right of Figure 8.2 where each idle place of the original net is split into a start place and an end one, the tokens in idle places are put into their corresponding start places, and processing times are added to the activity places. If an initial marking (e.g., all part tokens are in the start places) and a goal marking (e.g., all part tokens are in the end places) are given, the place-timed SC-net can be used to find a transition firing sequence with the lowest cost, which represents an optimal execution sequence of the underlying system.

8.2.3 Synthesis of a Place-Timed Petri Net

Untimed Petri nets can be constructed by either a top-down or bottom-up approach for a discrete event system (Jeng & DiCesare, 1993; Wu & Zhou, 2010;

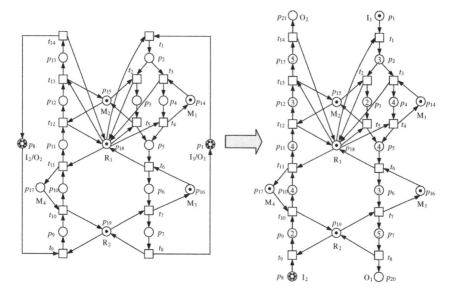

Figure 8.2 Conversion of an S^3PR net into an SC-net.

Zhou & DiCesare, 1993;), and so do place-timed SC-nets that are used for the RAS scheduling. In the top-down approach, first the high-level description of a system is presented and then the model is refined in a stepwise manner until a complete net model is achieved, whereas in the bottom-up approach, the net modules of subsystems are constructed, and then, they are combined by merging some common places, transitions, and subnets to form a whole model.

8.2.3.1 Top-down Method

The top-down synthesis method begins with an aggregate model of a system and neglects low-level details. Then, a refinement is done in a stepwise manner to incorporate more details into the model. Two schemes for the refinement can be used, i.e., expanding places and expanding transitions. The refinement continues until the level of details satisfies the specifications of the system. The top-down method has the advantage of starting with a global view of the system. However, if we want to preserve system properties, the top-down method is not easily applied to concurrent environments, such as RASs. The reason is that the interactions among the subsystems of these systems are coupled throughout all steps of synthesis, which makes it difficult to specify the system by using the top-down method. For such systems, the following bottom-up synthesis provides a complementary method.

8.2.3.2 Bottom-up Method

The bottom-up synthesis of place-timed SC-nets for RASs works as follows. First, each processing subnet is separately modeled, ignoring interactions with other subnets. These subnets are usually simple and easy to verify. Different subnets may have common places, e.g., the resource places shared by different activity places, which represent the interactions among the subnets. At each synthesis step, these interactions are considered and the processing subnets are combined through merging these places into a larger system. Finally, operation times are put on the corresponding activity places.

We use the manufacturing system given in Example 8.1 to illustrate the bottom-up synthesis method. First, we specify the individual processing subnets separately. At this stage, any interaction among the subnets is neglected. Figure 8.3a shows the processing of the part type of P_1 and the initial merges of the places for R_1, R_2, M_1, M_2, and M_3. Figure 8.3b shows the processing of P_2 and the initial merges of the places for R_1, R_2, M_2, and M_4. Then, a net model for the entire system is obtained as in Figure 8.3c by merging the two subnets through the resource places, i.e., the resource places with the same description are merged into a single place. Finally, after adding initial tokens and operation times into the net, the SC-net of the system is obtained as in Figure 8.3d. Note that a hybrid Petri net synthesis approach that blends top-down and bottom-up ideas is given in Zhou *et al.* (1992).

8.3 State Evolution of Place-Timed Nets

After an RAS is modeled by place-timed SC-nets, the dynamic behavior of the model can be described and analyzed by algebraic equations. This section presents the state evolution of SC-nets by firing sequence of enabled transitions.

Assume that activity places are κ-bounded in an SC-net. Let R be a $|P| \times \kappa$ remaining time matrix that indicates the remaining times of tokens in all places, $[N]^-$ and $[N]^+$ be the pre-incidence and post-incidence matrices of the net, D be the operation time vector associated with places, and C be the control vector that indicates which transition is to fire. In addition, $R(p_i, \cdot)$ denotes the row of R with respect to p_i, $R(p_i, k)$ represents the kth element in $R(p_i, \cdot)$, and M_+ and R_+ represent the marking vector and the remaining time matrix, respectively, after the removal of tokens in the pre-places of a transition t_{j*} that is selected to fire and before the addition of tokens to the post-places of t_{j*}. Since SC-nets have time information in activity places, a state of the underlying system should contain both the tokens' distributions M and the tokens' remaining time R. The following algorithm is presented to compute the immediate succeeding state (denoted by M' and R') of a given state (denoted by M and R) by firing a transition t_j^* enabled at M.

Input: An SC-net $\mathcal{N} = (P_S \cup P_E \cup P_R \cup P_A, T, F, W, D)$, a marking M, a remaining time matrix R, and a transition t_{j*} enabled at M.

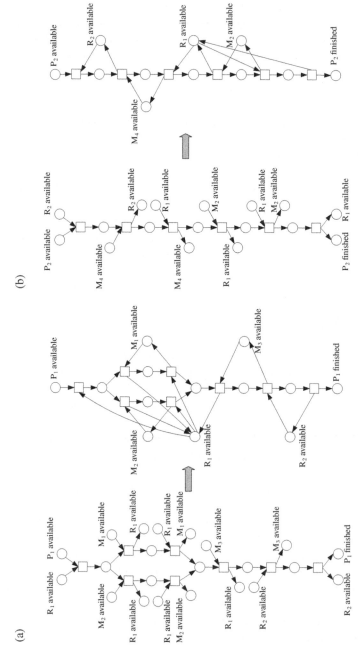

Figure 8.3 (a) Processing of P_1 (initial merges); (b) processing of P_2 (initial merges); (c) the model after merges; (d) the final SC-net with tokens and time delays.

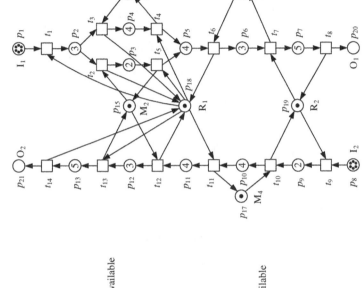

(d)

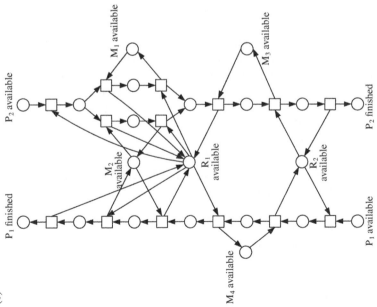

(c)

Figure 8.3 (Continued)

Output: An immediate succeeding state with M' and R' of a state with M and R by firing t_{j^*}.

1) The control vector C is set as follows: if $j = j^*$, $C(t_j) = 1$, otherwise $C(t_j) = 0$.
2) Remove tokens from the pre-places of t_{j^*}:

$$M_+ = M - [N]^- \cdot C. \tag{8.1}$$

3) Construct a set of t_{j^*}'s pre-places whose operation times are not zero, i.e., $P_{pre}(t_{j^*}) = \{p_i | p_i \in {}^\bullet t_{j^*} \wedge D(p_i) \neq 0\}$. If $P_{pre}(t_{j^*}) = \emptyset$, then $R_+ = R$; otherwise:
   3.1) Calculate the maximal remaining time among the tokens that are in $P_{pre}(t_{j^*})$ and required by the firing of t_{j^*} as follows:

$$\delta = \max_{\forall p_i \in P_{pre}(t_{j^*})} \left\{ R(p_i, M(p_i) - W(p_i, t_{j^*}) + 1) \right\}. \tag{8.2}$$

   3.2) Subtract δ from the remaining time matrix, i.e., $\forall p_i \in P$ and $\forall k \in \{1, \dots, \kappa\}$,

$$R_+(p_i, k) = \begin{cases} R(p_i, k) - \delta, & \text{if } R(p_i, k) \geq \delta; \\ 0, & \text{otherwise.} \end{cases} \tag{8.3}$$

4) Add new tokens to the post-places of t_{j^*}:

$$M' = M_+ + [N]^+ \cdot C. \tag{8.4}$$

5) Construct a set of t_{j^*}'s post-places whose operation times are not zero, i.e., $P_{post}(t_{j^*}) = \{p_i | p_i \in t_{j^*}^\bullet \wedge D(p_i) \neq 0\}$. If $P_{post}(t_{j^*}) \neq \emptyset$, then $\forall p_i \in P_{post}(t_{j^*})$, insert $D(p_i)$ into the first entry of $R_+(p_i, \cdot)$. Finally, $R' = R_+$.

Note that time delays are associated with activity places and the weight of any arc connecting an activity place is equal to one by Definition 8.2 for an SC-net. So, the values of $W(p_i, t_{j^*})$ and $W(t_{j^*}, p_i)$ in Steps 3.1 and 5 are equal to one for SC-nets. In addition, $\forall p_i \in P$, the entries in $R(p_i, \cdot)$ are ranked in a non-increasing magnitude of their values if $\kappa > 1$. The reason is that the remaining time of a new token in p_i is $D(p_i)$ which is biggest remaining time for the token and is inserted into the first entry of $R(p_i, \cdot)$.

We use the SC-net shown on the right of Figure 8.2 as an example to illustrate the procedures of state evolution of an SC-net. For this net, places are associated with the following operation times:

	p_1	p_2	p_3	p_4	p_5	p_6	p_7	p_8	p_9	p_{10}	p_{11}
	p_{12}	p_{13}	p_{14}	p_{15}	p_{16}	p_{17}	p_{18}	p_{19}	p_{20}	p_{21}	
$D^T =$	[0	3	2	4	4	3	5	0	2	4	4
	3	5	0	0	0	0	0	0	0	0]	

We see that all activity places of the net are 1-bounded, so $\kappa = 1$ and R is a $|P| \times 1$ matrix. The initial state of the net is:

	p_1	p_2	p_3	p_4	p_5	p_6	p_7	p_8	p_9	p_{10}	p_{11}
	p_{12}	p_{13}	p_{14}	p_{15}	p_{16}	p_{17}	p_{18}	p_{19}	p_{20}	p_{21}	
$M_0^T =$	[6	0	0	0	0	0	0	6	0	0	0
	0	0	1	1	1	1	1	1	0	0]	
$R_0^T =$	[0	0	0	0	0	0	0	0	0	0	0
	0	0	0	0	0	0	0	0	0	0]	

Given an SC-net $\mathcal{N} = (P_S \cup P_E \cup P_R \cup P_A, T, F, W, D)$, the pre-incidence matrix and post-incidence matrix of $\mathcal{N}$ can be given by the following expressions:

$$[N]^-(p, t) = \begin{cases} W(p, t), & \text{if } (p, t) \in F; \\ 0, & \text{otherwise.} \end{cases} \tag{8.5}$$

$$[N]^+(p, t) = \begin{cases} W(t, p), & \text{if } (t, p) \in F; \\ 0, & \text{otherwise.} \end{cases} \tag{8.6}$$

Thus, the pre-incidence matrix and post-incidence matrix of the given net are:

$$[N]^- = \begin{bmatrix}
1 & 0 & 0 & 0 & 0 & 0 & 0 & 0 & 0 & 0 & 0 & 0 & 0 & 0 \\
0 & 1 & 1 & 0 & 0 & 0 & 0 & 0 & 0 & 0 & 0 & 0 & 0 & 0 \\
0 & 0 & 0 & 0 & 1 & 0 & 0 & 0 & 0 & 0 & 0 & 0 & 0 & 0 \\
0 & 0 & 0 & 1 & 0 & 0 & 0 & 0 & 0 & 0 & 0 & 0 & 0 & 0 \\
0 & 0 & 0 & 0 & 0 & 1 & 0 & 0 & 0 & 0 & 0 & 0 & 0 & 0 \\
0 & 0 & 0 & 0 & 0 & 0 & 1 & 0 & 0 & 0 & 0 & 0 & 0 & 0 \\
0 & 0 & 0 & 0 & 0 & 0 & 0 & 1 & 0 & 0 & 0 & 0 & 0 & 0 \\
0 & 0 & 0 & 0 & 0 & 0 & 0 & 0 & 1 & 0 & 0 & 0 & 0 & 0 \\
0 & 0 & 0 & 0 & 0 & 0 & 0 & 0 & 0 & 1 & 0 & 0 & 0 & 0 \\
0 & 0 & 0 & 0 & 0 & 0 & 0 & 0 & 0 & 0 & 1 & 0 & 0 & 0 \\
0 & 0 & 0 & 0 & 0 & 0 & 0 & 0 & 0 & 0 & 0 & 1 & 0 & 0 \\
0 & 0 & 0 & 0 & 0 & 0 & 0 & 0 & 0 & 0 & 0 & 0 & 1 & 0 \\
0 & 0 & 0 & 0 & 0 & 0 & 0 & 0 & 0 & 0 & 0 & 0 & 0 & 1 \\
0 & 0 & 1 & 0 & 0 & 0 & 0 & 0 & 0 & 0 & 0 & 0 & 0 & 0 \\
0 & 1 & 0 & 0 & 0 & 0 & 0 & 0 & 0 & 0 & 0 & 1 & 0 & 0 \\
0 & 0 & 0 & 0 & 0 & 1 & 0 & 0 & 0 & 0 & 0 & 0 & 0 & 0 \\
0 & 0 & 0 & 0 & 0 & 0 & 0 & 0 & 0 & 1 & 0 & 0 & 0 & 0 \\
1 & 0 & 0 & 1 & 1 & 0 & 0 & 0 & 0 & 0 & 1 & 0 & 1 & 0 \\
0 & 0 & 0 & 0 & 0 & 0 & 1 & 0 & 1 & 0 & 0 & 0 & 0 & 0 \\
0 & 0 & 0 & 0 & 0 & 0 & 0 & 0 & 0 & 0 & 0 & 0 & 0 & 0 \\
0 & 0 & 0 & 0 & 0 & 0 & 0 & 0 & 0 & 0 & 0 & 0 & 0 & 0
\end{bmatrix}$$

$$[N]^+ = \begin{bmatrix}
0 & 0 & 0 & 0 & 0 & 0 & 0 & 0 & 0 & 0 & 0 & 0 & 0 & 0 \\
1 & 0 & 0 & 0 & 0 & 0 & 0 & 0 & 0 & 0 & 0 & 0 & 0 & 0 \\
0 & 1 & 0 & 0 & 0 & 0 & 0 & 0 & 0 & 0 & 0 & 0 & 0 & 0 \\
0 & 0 & 1 & 0 & 0 & 0 & 0 & 0 & 0 & 0 & 0 & 0 & 0 & 0 \\
0 & 0 & 0 & 1 & 1 & 0 & 0 & 0 & 0 & 0 & 0 & 0 & 0 & 0 \\
0 & 0 & 0 & 0 & 0 & 1 & 0 & 0 & 0 & 0 & 0 & 0 & 0 & 0 \\
0 & 0 & 0 & 0 & 0 & 0 & 1 & 0 & 0 & 0 & 0 & 0 & 0 & 0 \\
0 & 0 & 0 & 0 & 0 & 0 & 0 & 0 & 0 & 0 & 0 & 0 & 0 & 0 \\
0 & 0 & 0 & 0 & 0 & 0 & 0 & 0 & 1 & 0 & 0 & 0 & 0 & 0 \\
0 & 0 & 0 & 0 & 0 & 0 & 0 & 0 & 1 & 0 & 0 & 0 & 0 & 0 \\
0 & 0 & 0 & 0 & 0 & 0 & 0 & 0 & 0 & 1 & 0 & 0 & 0 & 0 \\
0 & 0 & 0 & 0 & 0 & 0 & 0 & 0 & 0 & 0 & 1 & 0 & 0 & 0 \\
0 & 0 & 0 & 0 & 0 & 0 & 0 & 0 & 0 & 0 & 0 & 1 & 0 & 0 \\
0 & 0 & 0 & 1 & 0 & 0 & 0 & 0 & 0 & 0 & 0 & 0 & 0 & 0 \\
0 & 0 & 0 & 0 & 1 & 0 & 0 & 0 & 0 & 0 & 0 & 0 & 1 & 0 \\
0 & 0 & 0 & 0 & 0 & 0 & 1 & 0 & 0 & 0 & 0 & 0 & 0 & 0 \\
0 & 0 & 0 & 0 & 0 & 0 & 0 & 0 & 0 & 0 & 1 & 0 & 0 & 0 \\
0 & 1 & 1 & 0 & 0 & 1 & 0 & 0 & 0 & 0 & 0 & 1 & 0 & 1 \\
0 & 0 & 0 & 0 & 0 & 0 & 0 & 1 & 0 & 1 & 0 & 0 & 0 & 0 \\
0 & 0 & 0 & 0 & 0 & 0 & 0 & 1 & 0 & 0 & 0 & 0 & 0 & 0 \\
0 & 0 & 0 & 0 & 0 & 0 & 0 & 0 & 0 & 0 & 0 & 0 & 0 & 1
\end{bmatrix}$$

Now we assume that transition t_1 which is enabled at the current marking is selected to fire. Then, the control vector C is:

	t_1	t_2	t_3	t_4	t_5	t_6	t_7	t_8	t_9	t_{10}	t_{11}	t_{12}	t_{13}	t_{14}
$C^T =$	[1	0	0	0	0	0	0	0	0	0	0	0	0	0]

By using the token movement equation in Equation (8.1), we have:

$$M_{0+} = M_0 - [N]^- \cdot C.$$

	p_1	p_2	p_3	p_4	p_5	p_6	p_7	p_8	p_9	p_{10}	p_{11}
	p_{12}	p_{13}	p_{14}	p_{15}	p_{16}	p_{17}	p_{18}	p_{19}	p_{20}	p_{21}	
$M_{0+}^T =$	[5	0	0	0	0	0	0	6	0	0	0
	0	0	1	1	1	1	0	1	0	0]	

Since $P_{pre}(t_1) = \emptyset$, we have

$$R_{0+} = R_0.$$

Then, after adding tokens to the post-places of t_1 by Equation (8.4), the marking vector M_1 is obtained as follows:

p_1	p_2	p_3	p_4	p_5	p_6	p_7	p_8	p_9	p_{10}	p_{11}
p_{12}	p_{13}	p_{14}	p_{15}	p_{16}	p_{17}	p_{18}	p_{19}	p_{20}	p_{21}	

$M_1^T =$

[5	1	0	0	0	0	0	6	0	0	0
0	0	1	1	1	1	0	1	0	0]	

Since $P_{post}(t_1) = \{p_2\}$ and $D(p_2) = 3$, we insert 3 into the first entry of $R_{0+}(p_2, \cdot)$ and obtain R_1 as described below:

p_1	p_2	p_3	p_4	p_5	p_6	p_7	p_8	p_9	p_{10}	p_{11}
p_{12}	p_{13}	p_{14}	p_{15}	p_{16}	p_{17}	p_{18}	p_{19}	p_{20}	p_{21}	

$R_1^T =$

[0	3	0	0	0	0	0	0	0	0	0
0	0	0	0	0	0	0	0	0	0]	

Next, we suppose that t_3 is selected to fire at the current state. Then, the control vector C is:

t_1	t_2	t_3	t_4	t_5	t_6	t_7	t_8	t_9	t_{10}	t_{11}	t_{12}	t_{13}	t_{14}

$C^T =$ [0 0 1 0 0 0 0 0 0 0 0 0 0 0]

By Equation (8.1), the marking vector M_{1+} after the token removing is:

p_1	p_2	p_3	p_4	p_5	p_6	p_7	p_8	p_9	p_{10}	p_{11}
p_{12}	p_{13}	p_{14}	p_{15}	p_{16}	p_{17}	p_{18}	p_{19}	p_{20}	p_{21}	

$M_{1+}^T =$

[5	0	0	0	0	0	0	6	0	0	0
0	0	0	1	1	1	0	1	0	0]	

For the net, we have $P_{pre}(t_3) = \{p_2\}$. According to Equation (8.2), $\delta = \max\{R(p_2, 1)\} = 3$. Then, by following Equation (8.3), we have:

p_1	p_2	p_3	p_4	p_5	p_6	p_7	p_8	p_9	p_{10}	p_{11}
p_{12}	p_{13}	p_{14}	p_{15}	p_{16}	p_{17}	p_{18}	p_{19}	p_{20}	p_{21}	

$R_{1+}^T =$

[0	0	0	0	0	0	0	0	0	0	0
0	0	0	0	0	0	0	0	0	0]	

After adding new tokens to the post-places of t_3 by Equation (8.4), the marking vector M_2 is:

	p_1	p_2	p_3	p_4	p_5	p_6	p_7	p_8	p_9	p_{10}	p_{11}
	p_{12}	p_{13}	p_{14}	p_{15}	p_{16}	p_{17}	p_{18}	p_{19}	p_{20}	p_{21}	
$M_2^T =$	[5	0	0	1	0	0	0	6	0	0	0
	0	0	0	1	1	1	1	1	0	0]	

We have that $P_{post}(t_3) = \{p_4\}$ and $D(p_4) = 4$. So, we insert 4 into the first entry of $R_{1+}(p_4, \cdot)$ and obtain R_2 as:

	p_1	p_2	p_3	p_4	p_5	p_6	p_7	p_8	p_9	p_{10}	p_{11}
	p_{12}	p_{13}	p_{14}	p_{15}	p_{16}	p_{17}	p_{18}	p_{19}	p_{20}	p_{21}	
$R_2^T =$	[0	0	0	4	0	0	0	0	0	0	0
	0	0	0	0	0	0	0	0	0	0]	

The state evolution of an SC-net continues like this and a sequence of reachable states are generated by firing a sequence of transitions.

8.4 A* Search on a Reachability Graph

This section introduces a heuristic search method which applies the well-known A* algorithm within the reachability graphs of place-timed Petri nets to find optimized scheduling paths. This method does not need to explore the whole reachability graph of the net to obtain a schedule from an initial marking to a given goal marking. In addition, if its used heuristic function is admissible, this method guarantees to obtain an optimal schedule for the underlying system. The A* search within a Petri net reachability graph is given in Algorithm 8.1.

Algorithm 8.1

```
A* search within a reachability graph:
    1) Put the initial marking M₀ on the list OPEN.
    2) If OPEN is empty, terminate with failure.
    3) Remove the first marking M from OPEN and put it on the list
CLOSED.
    4) If M equals a goal marking M_G, construct the schedul-
ing path from M₀ to M_G and terminate.
    5) Find the set of enabled transitions of M.
    6)  For each transition t enabled at M, generate an immediate
successor M′ by firing t at M, set a pointer from M′ to M, and
```

compute $g(M') = g(M) + c(M, M')$, $h(M')$, and $f(M') = g(M') + h(M')$.

    a) If M' is already on OPEN, direct its pointer along the path yielding the smallest $g(M')$.

    b) If M' is already on CLOSED, direct its pointer along the path yielding the smallest $g(M')$. If M' requires such a pointer redirection, move M' to OPEN.

    c) If M' is on neither OPEN nor CLOSED, put M' on OPEN.

  7) Reorder OPEN in the non-decreasing magnitude of the f-values.

  8) Go to Step 2).

In Algorithm 8.1, $f(M) = g(M) + h(M)$ is an estimate of the lowest cost $f^*(M)$ from M_0 to M_G among all paths going through the current marking M, $g(M)$ is the current lowest cost obtained from M_0 to M, $h(M)$ is an estimate of the lowest cost $h^*(M)$ from M to M_G among all paths, and $c(M, M')$ denotes the cost of an edge (M, M') in the reachability graph of the net. The OPEN list contains the markings that have been generated but not expanded and the CLOSED list contains the markings that have been expanded. Algorithm 8.1 is an A*-based algorithm. So its properties, such as optimality and completeness, are the same as the properties of the A* algorithm in Section 2.3.2. Note that if more than one marking has the smallest f-value in OPEN, a tie-breaking rule must be applied to choose a marking for expansion. The last-in-first-out rule is used as the tie-breaking rule in this book.

8.5 A* Search with State Check

However, Algorithm 8.1 cannot be directly applied to the reachability graphs of SC-nets. In Algorithm 8.1, if a new marking M' is generated as an immediate successor of a marking M, it is compared with all markings in OPEN and CLOSED to check whether there already exists a same marking as M' or not. If such a marking exists, its g-value is compared with the g-value of M' to decide if the path to the marking is more promising than the path to M'. However, for an SC-net, more than one state may have the same marking but different tokens' remaining time matrices, which represent different states of the underlying system. So, Algorithm 8.1, which does not consider the tokens' remaining times, may prune states that exist in paths leading to optimal solutions of the SC-net.

To fix the problem, the states of an SC-net should be correctly defined. In an SC-net, a token in a timed place p may mean that i) this token is already available for the firing of a post-transition of p; or ii) although the token is not ready yet, it should become available after a fixed time. Note that, in the second case, nonzero remaining time exists for the token to be available. As time passes by, the token's remaining time is decreased until it reaches zero. When it reaches zero, the token becomes available to fire a post-transition of its place. Thus, the

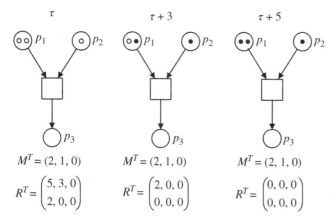

Figure 8.4 Three states with the same marking but different tokens' remaining time matrices.

tokens' remaining times are just as important as the tokens' distribution for an SC-net to "reach" a specific state of the underlying system. Therefore, we need both the marking and the tokens' remaining time matrix to denote a state in SC-nets.

Definition 8.3 *Given a place-timed SC-net $\mathcal{N}$ whose activity places are κ-bounded. A state of $\mathcal{N}$ is defined as $S = (M, R)$ where M denotes a marking and R denotes a $|P| \times \kappa$ tokens' remaining time matrix that represents the remaining time of each token in all places. S_0 is called an initial state and $(\mathcal{N}, S_0)$ is called a timed net system of $\mathcal{N}$.*

For example, consider three states of a simple place-timed net shown in Figure 8.4 where circles in places denote unavailable tokens and black dots represent available tokens. Suppose that p_1 and p_2 are timed places that are 2-bounded. We can see that all the three states have the same marking M, but they are actually different states at time τ, $\tau + 3$, and $\tau + 5$ since they have different tokens' remaining time matrices. So, they do not represent the same state of the system. For the place-timed net on the left of Figure 8.4, although the transition is enabled at the tokens' distribution M, it is not enabled at the state S since the tokens required by the firing of the transition are not all available. The transitions in the other two nets are enabled at their state in terms of both tokens' distribution M and tokens' remaining times R.

To schedule SC-nets, we present the A* algorithm that considers the state check and both the tokens' distribution M and the tokens' remaining times R as in Algorithm 8.2.

Algorithm 8.2

```
A* search with a state check for SC-nets:
    1) Place the initial state S₀ on the list OPEN.
    2) If OPEN is empty, terminate with failure.
    3) Remove the first state S = (M, R) from OPEN and put S on the
list CLOSED.
    4) If S equals a goal state S_G, construct the path from S back
to S₀ and terminate.
    5) Find the set of transitions enabled at M.
    6) For each transition t enabled at M, decrease the tokens'
remaining times to enable t at both M and the tokens' remaining
times, generate an immediate succeeding state S' = (M', R') by firing
t, set a pointer from S' to S, and calculate g(S') = g(S) + c(S, S'), h(S'),
and f(S') = g(S') + h(S').
        a) If there exists a state S^O = (M^O, R^O) on OPEN such that
M' = M^O, R' = R^O, and g(S') < g(S^O), delete S^O from OPEN and insert
S' on OPEN.
        b) If there exists a state S^C = (M^C, R^C) on CLOSED such that
M' = M^C and R' = R^C, and g(S') < g(S^C), delete S^C from CLOSED and
insert S' on OPEN.
        c) If both lists do not contain a state with M' and R',
insert S' on OPEN.
    7) Reorder OPEN in the non-decreasing magnitude of f-values.
    8) Go to Step 2).
```

8.6 An Illustrative Example

In this section, the model on the right of Figure 8.2 is tested as an example. Suppose that the initial state has one token in both p_1 and p_8 and the goal state has one token in both p_{20} and p_{21}. We use $h = 0$ as a heuristic function for the A* search within the reachability graph. For this net, the number of expanded states is 99 and its computational time is less than 1 second. The obtained system schedule is shown in Table 8.1 in the form of a firing sequence of transitions. The makespan of the obtained schedule is 21, which is guaranteed to be minimal since the used heuristic function is admissible. For the evolution and scheduling of SC-nets with

Table 8.1 The obtained schedule for the SC-net on the right of Figure 8.2 with $M_0(p_1) = M_0(p_8) = 1$ (makespan = 21).

Transition	Fire time	Transition	Fire time	Transition	Fire time
t_9	0	t_5	5	t_{12}	13
t_1	0	t_6	9	t_{13}	16
t_{10}	2	t_{11}	9	t_8	17
t_2	3	t_7	12	t_{14}	21

the A* search and state checks presented in this chapter, we have developed some C# programs. The source codes of the presented methods and all experimental files are available in (Huang, 2019a) for a reader's reference.

8.7 Concluding Remarks

First, this chapter presents the definitions of place-timed SC-nets for RASs and two kinds of modeling methods for RASs based on SC-nets: conversion from existing untimed models and synthesis via a top-down or bottom-up method. Then, the state evolution, which considers both the tokens' distribution and the tokens' remaining times, is presented for such nets. Finally, the method that combines Petri net evolution with A* search algorithm is introduced to obtain executable schedules for the underlying systems of SC-nets. The state check rules for such models are also given. The approach can be applied to the scheduling of RASs with SC-nets. In addition, it only needs to explore a partial reachability graph with a heuristic function to obtain a feasible schedule and the schedule is guaranteed to be optimal if the used heuristic function is admissible.

However, like other A* based method, the presented method suffers from the computational complexity problem since the number of explored markings grows exponentially with the problem size and/or initial markings. The following chapters introduce different strategies that can be used to accelerate the search process, making the approach applicable to larger and more complex RASs.

8.8 Bibliographical Notes

The classifications and applications of timed Petri nets can be found in the book (Wang, 1998). For the modeling methods based on Petri nets, please refer to the books (Hrúz & Zhou, 2007; Moody & Antsaklis, 2012; Seatzu *et al.*, 2012; Zhou & DiCesare, 1993). Tuncel and Bayhan (2007) provide a comprehensive review of the research work on the scheduling methods with Petri nets. Rios and Chaimowicz (2010) give a survey and classification of A*-based heuristic search algorithm. The complexity analysis of the A* algorithm can be seen in Korf and Reid (1998) and Russell and Norvig (2010). For more details about the heuristic search algorithms, please refer to Edelkamp and Schroedl (2012) and Pearl (1984). The material presented in this chapter on the state check mainly comes from Huang and Sun (2005). Some recent progress along this direction can be found in Luo *et al.*, (2015).

9

Controllable Heuristic Search

This chapter presents and evaluates two heuristic search strategies and their applications to the scheduling of resource allocation systems (RASs) in an SC-net framework. SC-nets can concisely model strict precedence constraints, alternative routes, multiple kinds of resources, concurrent activities, and operation times, which are common in RASs. To handle the schedule complexities of the SC-nets, this chapter presents an admissible heuristic function for these nets and two controllable heuristic A* search methods by using a heuristic function. The first search method does not need to predict the depth of a solution in advance and the quality of the obtained schedule is controllable. The second search method combines an A* search with a depth-first strategy based on the execution of SC-nets to cope with the complexities of RAS scheduling. It can invoke quicker termination conditions and the quality of the obtained schedule is controllable. Some numerical experiments are carried out to demonstrate the effectiveness and efficiency of the methods.

9.1 Introduction

Resource allocation systems (RASs) are a kind of discrete event dynamic systems that have multiple concurrent flows of processes. Different products such as parts, vehicles, programs, and data are concurrently processed in the systems and some shared resources such as machines, robots, and drives are exploited to reduce the systems' cost. In RASs, for instance, automated manufacturing systems, project management systems, and software engineering systems, there often exists a high-level control agent to decide which resource is assigned to which process and at what time to optimize some criteria, e.g., makespan and tardiness. Thus, the purpose of RAS scheduling is to determine when to perform which activity by which resource such that the system constraints are met and the system objectives are best achieved.

Supervisory Control and Scheduling of Resource Allocation Systems: Reachability Graph Perspective, First Edition. Bo Huang and MengChu Zhou.
© 2020 The Institute of Electrical and Electronics Engineers, Inc. Published 2020 by John Wiley & Sons, Inc.

Many industrial and research communities focus on developing some methods to quickly and optimally solve RAS scheduling problems, which is a challenge because no perfect solution has been in existence for all problems, due primarily to the complexity of RAS scheduling. For example, the scheduling of flexible manufacturing systems belongs to NP-hard combinatorial problems (Tuncel & Bayhan, 2007) for which it is unlikely to develop an optimal algorithm with polynomial complexity in general.

RAS scheduling has also been studied by the Petri net community since Petri nets are a mathematical formalism and a graphical tool that is suitable for the modeling and analysis of RASs. As a graphical tool, Petri nets work like flow charts and provide a visualization of discrete-event dynamic systems. As a mathematical tool, they can be described by a set of linear algebraic equations which allow a formal check on the behavioral properties of the underlying systems. A method that combines Petri net simulation capabilities with the A* search algorithm within reachability graphs is pioneered by Lee and DiCesare (1994) and then applied in Baruwa *et al.* (2015), Huang *et al.* (2008, 2014), Lee and Lee (2010), Luo *et al.* (2015), Mejía and Odrey (2005), Mejía *et al.* (2018), Moro *et al.* (2002b), and Xiong and Zhou (1998). The method has great merit that once the Petri net model of a system is constructed, an optimal schedule from a given initial state to a given goal state can be found without exploring all reachable states. Moreover, if an admissible heuristic function is used, the method terminates with an optimal schedule.

However, there are also some problems in the method. First, the often used admissible heuristic function (Huang & Sun, 2005; Huang *et al.*, 2010; Xiong *et al.*, 1996; Xiong & Zhou, 1998) is only for the RASs without alternative routes. In fact, alternative routes are common in RASs. Second, for large and complex systems, it is usually difficult for an A* search algorithm with an admissible heuristic function to terminate with a solution in a reasonable amount of time.

To accelerate a search process, some improved algorithms have been proposed. For example, Xiong and Zhou (1998) propose two hybrid algorithms that combine Best-First with Backtracking search methodologies. The A*-based algorithms that limit backtracking capability by introducing irrevocable decisions are implemented in Inaba *et al.* (1998), Jeng *et al.* (1998), and Mejía (2002). Hybrid A*-based algorithms with relaxation of an evaluation scope have been proposed in Baruwa *et al.* (2015), Lee and Lee (2010), Luo *et al.* (2015), Moro *et al.* (2002b), and Yu *et al.* (2003). Although these methods can speed up the A* search within reachability graphs, the quality of obtained solutions is uncontrollable, i.e., the deviations of the obtained costs from the optimal cost are unpredictable. To control the quality of the obtained schedules, Pohl (1973) proposes a dynamic weighted A* (DWA) algorithm to cut down the computational time at the expense of narrowing down an evaluation scope, thus risking missing optimal schedules. The DWA algorithm with an admissible heuristic function can obtain a solution whose cost does not exceed the optimal cost by more than a factor $1 + \varepsilon (\varepsilon \geq 0)$. However, it needs to estimate the depth of the solution in advance and it is thus inapplicable to the RASs with alternative routes since they have solutions with different depths.

To address these problems, this chapter first presents an admissible heuristic function that is suitable for the scheduling of SC-nets for RASs with alternative routes. Then, two different controllable A* search methods with a heuristic function are presented to schedule these models. The first method does not need to anticipate the depth of a solution and the quality of the search result is controllable. The second method combines A* search with a depth-first strategy that invokes quicker termination conditions and the quality of the obtained result is also controllable.

9.2 Alternative Routes with Different Lengths

Alternative routes exist in many RASs and they indicate that some tasks in RASs can be performed by different processing routes of the systems. As an example, this section presents an automated manufacturing system with shared resources and alternative routes. The system consists of four jobs and the job requirements are shown in Table 9.1 in which the required time of each operation is given in parentheses. The system has three types of shared resources (R_1, R_2, and R_3) and four jobs (P_1, P_2, P_3, and P_4). Each job has three tasks (T_1, T_2, and T_3) and some tasks have alternative routes of operations. For example, T_3 of P_1 can be performed with either resource R_1 or resources R_1 and R_3, while the corresponding processing times are 80 and 108 (57 + 51), respectively. Between any two sequential operations of the system, there is an intermediate buffer that stores parts for the next operation. The SC-net of the system is given in Figure 9.1 where the activity places without an integer number denote the intermediate

Table 9.1 Job requirements of the example system.

	P_1	P_2	P_3	P_4
T_1	$R_3(69)$ or $R_2(75)$	$R_2(95)$	$R_2(78)$	$R_3(99)$
T_2	$R_3(85)$	$R_2(85)$ or $R_3(97)$	$R_3(75)$	$R_3(76)$
T_3	$R_1(80)$ or $[R_1(57)$ and $R_3(51)]$	$R_1(98)$ or $R_3(92)$	$R_1(68)$ or $[R_3(56)$ and $R_2(70)]$	$R_1(93)$

Source: Adapted from Huang *et al.* (2012*a*). Reproduced with permission of Springer.

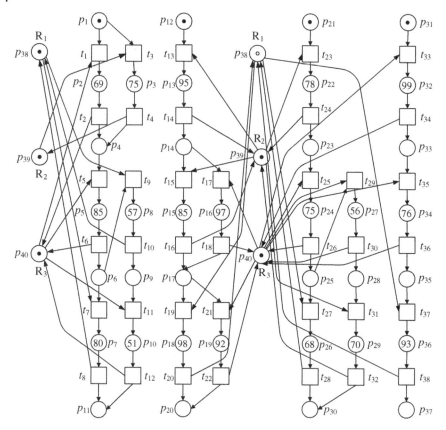

Figure 9.1 An SC-net of an automated manufacturing system. Source: Adapted from Huang *et al.* (2012*a*). Reproduced with permission of Springer.

buffers and the resource places with the same name represent the same type of resources.

9.3 An Admissible Heuristic for SC-nets

In an SC-net, firing an enabled transition changes tokens' distribution (i.e., marking M) and tokens' remaining time matrix R, which are represented by a state $S = (M, R)$. Firing a sequence of transitions thus generates a sequence of reachable states. All possible behaviors of the underlying system can be completely tracked by all reachable states that are contained in the reachability graph of the net. Therefore, the scheduling problem of an SC-net becomes finding a transition firing sequence in the reachability graph from the initial state S_0 to a given goal state S_G that is usually the one where all tokens in non-resource places are moved into their end places. To efficiently find such a scheduling path, an A* search algorithm can be used with a heuristic function h that evaluates each generated state S, i.e., $f(S) = g(S) + h(S)$ where $f(S)$ is an estimate of the cost from S_0 to S_G along an optimal path which goes through S, $g(M)$ is the current lowest cost obtained from S_0 to the current one S, and $h(S)$ is an estimate of the cost from S to S_G. The advantages of the A* algorithm are that it does not need to explore all the reachable states in the reachability graph to obtain a solution. If h is admissible, i.e., $\forall S, h(S) \le h^*(S)$ where $h^*(S)$ is the lowest cost of paths going from S to S_G, the algorithm terminates with an optimal solution.

In RAS scheduling, many A*-based methods (Huang & Sun, 2005; Huang *et al.*, 2010; Xiong & Zhou, 1998; Xiong *et al.*, 1996) use the following efficient and admissible heuristic function:

$$h(M) = \max_i \left\{ \xi_i(M), i = 1, 2, \dots, |P_R| \right\} \tag{9.1}$$

where $\xi_i(M)$ is the total operation time of the remaining operations that need to be performed with the ith resource from M to the goal marking. However, the heuristic function cannot be applied to all SC-nets, for example, the SC-net given in Figure 9.1. This is because the function does not consider alternative routes, weighted arcs, multiple copies of a type of resource, or the tokens' remaining times that are common in the SC-nets of RASs. This section presents a more general heuristic function that can be applied to such SC-nets. In addition, the proof of its admissibility is also given.

Before introducing the heuristic function, we first give the definitions of weighted operation time (WOT) and weighted resource time (WRT). WOT is defined as a $|P_A| \times |P_R|$ matrix for an SC-net. If an activity place p_i requires a kind of resource r, $\text{WOT}(p_i, r)$ denotes the WOT of an available token in p_i with r when all units of r are assumed to be concurrently used by the operations that require r, i.e., $\text{WOT}(p_i, r) = D(p_i) \times \frac{U_{p_i}(r)}{M_0(r)}$ where $U_{p_i}(r)$ denotes the units of r that are required by p_i. We use the SC-net in Figure 9.1 as an example. For this net, the activity place p_2 uses the resource $R_3 (p_{40})$ and its operation time $D(p_2)$ is 69. Since $U_{p_2}(R_3) = 1$ and $M_0(R_3) = 1$, we have that $\text{WOT}(p_2, p_{40}) = 69$. Based on WOT, WRT is defined as a $|P \backslash P_R| \times |P_R|$ matrix such that $\forall p_i \in P \backslash P_R, \forall r \in P_R$, $\text{WRT}(p_i, r)$ denotes the minimal total time of WOTs that are required by the activity places with r in a path for an available token in p_i to reach its end place. For example, the WRT matrix of Figure 9.1 is given as follows:

$$
\text{WRT} =
\begin{bmatrix}
p_1 : & 57 & 0 & 85; & p_{21} : & 0 & 78 & 75; \\
p_2 : & 57 & 0 & 85; & p_{22} : & 0 & 0 & 75; \\
p_3 : & 57 & 0 & 85; & p_{23} : & 0 & 0 & 75; \\
p_4 : & 57 & 0 & 85; & p_{24} : & 0 & 0 & 0; \\
p_5 : & 57 & 0 & 0; & p_{25} : & 0 & 0 & 0; \\
p_6 : & 57 & 0 & 0; & p_{26} : & 0 & 0 & 0; \\
p_7 : & 0 & 0 & 0; & p_{27} : & 0 & 70 & 0; \\
p_8 : & 0 & 0 & 51; & p_{28} : & 0 & 70 & 0; \\
p_9 : & 0 & 0 & 51; & p_{29} : & 0 & 0 & 0; \\
p_{10} : & 0 & 0 & 0; & p_{30} : & 0 & 0 & 0; \\
p_{11} : & 0 & 0 & 0; & p_{31} : & 93 & 0 & 175; \\
p_{12} : & 0 & 95 & 0; & p_{32} : & 93 & 0 & 76; \\
p_{13} : & 0 & 0 & 0; & p_{33} : & 93 & 0 & 76; \\
p_{14} : & 0 & 0 & 0; & p_{34} : & 93 & 0 & 0; \\
p_{15} : & 0 & 0 & 0; & p_{35} : & 93 & 0 & 0; \\
p_{16} : & 0 & 0 & 0; & p_{36} : & 0 & 0 & 0; \\
p_{17} : & 0 & 0 & 0; & p_{37} : & 0 & 0 & 0; \\
p_{18} : & 0 & 0 & 0; \\
p_{19} : & 0 & 0 & 0; \\
p_{20} : & 0 & 0 & 0;
\end{bmatrix}
$$

Note that $\text{WRT}(p_i, r)$ denotes the minimal time for an available token in p_i to reach its end place with resource r. As an example, for an available token in p_1, among all possible paths from p_1 to its end place p_{11}, it needs at least 57 units of time with resource R_1 (e.g., using a transition firing sequence $t_1 - t_2 - t_5 - t_6 - t_9 - t_{10} - t_{11} - t_{12}$). Similarly, it needs at least zero units of time with R_2 (e.g., with a firing sequence $t_1 - t_2 - t_5 - t_6 - t_7 - t_8$) and at least 85 units of time with R_3 (e.g., using a firing sequence $t_3 - t_4 - t_5 - t_6 - t_7 - t_8$). So, $\text{WRT}(p_1, \cdot) = (57, 0, 85)$. Suppose that all activity places are κ-bounded. A heuristic function for SC-nets can be given as:

$$
h_H(S) = \max_{r \in P_R} \left\{ \sum_{p_i \in P \backslash P_R} \left[M(p_i) \cdot \text{WRT}(p_i, r) + \sum_{x \in \mathbb{N}_\kappa} R(p_i, x) \cdot \frac{U_{p_i}(r)}{M_0(r)} \right] \right\} \quad (9.2)
$$

where $\mathbb{N}_\kappa = \{1, \ldots, \kappa\}$. In Equation (9.2), $M(p_i) \cdot \text{WRT}(p_i, r)$ denotes the total WRT with r for all tokens in p_i at S to reach their end place if the tokens are supposed to be available. $\sum_{x \in \mathbb{N}_\kappa} R(p_i, x) \cdot U_{p_i}(r)/M_0(r)$ denotes the remaining WRT with r for all tokens in p_i at S to be available. Note that in the computation of WRT with r, all units of r are assumed to be concurrently used by the operations that require r. Thus, $\sum_{p_i \in P \backslash P_R} \left[M(p_i) \cdot \text{WRT}(p_i, r) + \sum_{x \in \mathbb{N}_\kappa} R(p_i, x) \cdot U_{p_i}(r)/M_0(r) \right]$ represents the total WRT with r for all tokens in non-resource places at S to reach their end places. Therefore, $h_H(S)$ denotes the maximal total WRT with any resource for the system transferred from S to S_G. For instance, we consider the initial state $S_0 = (M_0, R_0)$ of the SC-net in Figure 9.1. Suppose that its goal sate S_G is the one where all non-resource tokens are in end places. Since $M_0(p_1) = M_0(p_{12}) = M_0(p_{21}) = M_0(p_{31}) = 1$ and $\forall r \in P_R$, $U_{p_1}(r) = U_{p_{12}}(r) = U_{p_{21}}(r) = U_{p_{31}}(r) = 0$, we have that $h_H(S_0) = \max\{57 \times 1 + 0 \times 1 + 0 \times 1 + 93, 0 \times 1 + 95 \times 1 + 78 \times 1 + 0 \times 1, 85 \times 1 + 0 \times 1 + 75 \times 1 + 175 \times 1\} = 335$, i.e., the maximal total WRT required by the system transferred from S_0 to S_G is 335 with R_3.

The formulation of h_H is motivated by the fact that a lower bound of the cost for the system transferred from the current state to a goal state can be calculated as the maximal weighted usage of any resource. The heuristic function h_H can be used for the SC-nets that have alternative routes, weighted arcs, and multiple copies of resources. In addition, the tokens' remaining processing time is also considered in the heuristic calculation, which makes h_H informed. In the following, we prove that it is admissible.

Theorem 9.1 h_H as defined in (9.2) is admissible for SC-nets.

Proof: According to the definition of h_H, for any reachable state S of an SC-net, $h_H(S)$ represents the maximal total WRT for all tokens in non-resource places at S to reach their end places with a resource r. It is calculated under the assumption that all units of r are concurrently used. In fact, the units of r are usually not concurrently used by operations that require them. In addition, there may exist some operations that use other types of resources and they also consume time. So, $h_H(S)$ is a lower bound of the cost or time needed by the system transferred from S to S_G. That is to say, $\forall S$, $h_H(S) \leq h^*(S)$ where $h^*(S)$ denotes the lowest cost among all possible paths from S to S_G. Therefore, h_H is admissible. ∎

9.4 A Controllable Heuristic Search

Although the A* algorithm can obtain an optimal result with an admissible heuristic function if such a result exists, it often spends a large amount of time discriminating among paths whose costs do not vary significantly from each other. In such a case, its admissibility property becomes a curse rather than a virtue. In some practical scenarios, we may be willing to accept a suboptimal solution in return for reduced computational time. In Pohl (1973), the DWA method is proposed to address this problem by adding additional dynamic weight to the heuristic function h to evaluate each generated node n as follows:

$$f(n) = g(n) + h(n) + \varepsilon \left[1 - \frac{d(n)}{d(n_G)} \right] h(n) \tag{9.3}$$

where $d(n)$ denotes the depth of node n and n_G represents the given goal node. The DWA method requires an estimate of the depth of all solutions and then decreases $1 - d(n)/d(n_G)$ from 1 at the initial node to 0 (or nearly 0) at the estimated goal depth. This maintains ε-admissibility, that is, if h is admissible, the algorithm finds a path with the cost at most $(1 + \varepsilon)c^*$ in which c^* represents the lowest cost of paths going from the initial node to the goal node. However, DWA is inapplicable to some systems (e.g., the systems with alternative routes) since the depth of the optimal solution, or at least a good upper bound to it, must be known a priori.

To handle it, this section presents an improved dynamic weighted A* algorithm (IDWA) which has the following advantages. First, the dependency on the notion of an anticipated depth of the solution is avoided. Second, if an admissible heuristic function is used, the quality of the obtained result is controllable, that is, it always finds a solution whose cost does not exceed the optimal cost by more than a factor $1 + \varepsilon$. The IDWA algorithm for SC-nets is implemented in Algorithm 9.1.

Algorithm 9.1

```
Algorithm IDWA {
1)      Set a desired value of ε such that ε ≥ 0.
2)      Put the initial state S₀ on the list OPEN.
3)      If OPEN is empty, terminate with failure.
4)      Remove the first state S = (M, R) from OPEN and put S on the
list CLOSED.
5)      If S equals the given goal state S_G, construct the path from
S₀ to S_G and terminate.
6)      Find the set of transitions enabled at M.
7) For each transition t enabled at M, decrease the tokens' remain-
ing times to enable t with respect to both M and R, generate its
immediate succeeding state S' = (M', R') by firing t, set a pointer from
S' to S, and calculate g(S') = g(S) + c(S, S'), h(S'), and f(S') = g(S') + h(S') +
ε (h(S'))/(h(S₀)) h(S').
        a) If there exists a state S^O = (M^O, R^O) on OPEN such that M' =
M^O, R' = R^O, and g(S') < g(S^O), delete S^O from OPEN and insert S' on
OPEN.
        b) If there exists a state S^C = (M^C, R^C) on CLOSED such that M' =
M^C and R' = R^C, and g(S') < g(S^C), delete S^C from CLOSED and insert S' on
OPEN.
        c) If both lists do not contain a state with M' and R', insert S'
on OPEN.
8) Reorder OPEN in the non-decreasing magnitude of f-values.
9)      Go to Step 3).
}
```

Note that IDWA uses $f(S') = g(S') + h(S') + \varepsilon \frac{h(S')}{h(S_0)} h(S')$ to evaluate each generated state. At shallow levels of the search space of IDWA, h is given a supportive weight which is greater than one, encouraging depth-first excursions. However, at deep levels where termination is likely to occur, the search assumes an admissible heuristic that is equal to h to resist possible premature termination. Next, we prove that if the used heuristic h is admissible, the IDWA algorithm is ε-admissible, i.e., it finds a path with the cost of $(1 + \varepsilon)c^*$ at most.

Theorem 9.2 *IDWA is ε-admissible if its used heuristic h is admissible.*

Proof: At each iteration of IDWA, a state with the minimal f-value in OPEN is selected for expansion. Let such a state be S'. If S' equals S_G, the algorithm terminates with a path from S_0 to S_G; otherwise, it is expanded to generate its immediate succeeding successors. According to Algorithm 9.1, we have:

$$f(S') = g(S') + h(S') + \varepsilon \frac{h(S')}{h(S_0)} h(S'). \tag{9.4}$$

Let $g^*(S')$ be the lowest cost among all paths from S_0 to S'. Since h is admissible and S' has the minimal f-value in OPEN, we have that $g(S') = g^*(S')$ (Pearl, 1984). Therefore:

$$f(S') \leq g^*(S') + h^*(S') + \varepsilon \frac{h(S')}{h(S_0)} h^*(S')$$
$$\leq c^* + \varepsilon h^*(S')$$
$$\leq (1 + \varepsilon)c^* \qquad\blacksquare$$

For the aforementioned example, we use the IDWA algorithm and the A* algorithm with the same admissible heuristic function h_H to show the effectiveness and efficiency of IDWA. Let c be the makespan of the obtained system schedule and N_E be the number of expanded states. The percentage of lost optimality $\mathcal{R}_c$, which is a comparison of the makespan obtained by IDWA with the makespan obtained by the A* algorithm, is defined as

$$\mathcal{R}_c = \frac{c(\text{IDWA}) - c(\text{A}^*)}{c(\text{A}^*)} \times 100\% \tag{9.5}$$

and the percentage of the reduced search effort $\mathcal{R}_E$, which is a comparison of the number of expanded states of IDWA with that of the A* algorithm, is defined as

$$\mathcal{R}_E = \frac{N_E(\text{IDWA}) - N_E(\text{A}^*)}{N_E(\text{A}^*)} \times 100\%. \tag{9.6}$$

The scheduling results with respect to the makespan, the percentage of lost optimality, the number of expanded states, and the percentage of reduced search effort are shown in Table 9.2. The optimal makespan and the number of expanded states of the A* algorithm, i.e., IDWA with $\varepsilon = 0$, are also given in the table.

From Table 9.2, we can see that IDWA yields a sharp drop in search effort when compared with the A* algorithm (i.e., $\mathcal{R}_E$ is low) while the solution found at termination is not worse than the optimal one by a factor greater than $1 + \varepsilon$ (i.e., $\mathcal{R}_c \leq \varepsilon$).

Table 9.2 Scheduling results of the example system.

ε	c	$\mathcal{R}_c$(%)	N_E	$\mathcal{R}_E$(%)
0.00	427	0.00	1442	0.00
0.05	427	0.00	688	−52.29
0.10	427	0.00	662	−54.09
0.15	435	1.87	638	−55.76
0.20	435	1.87	596	−58.67
0.25	435	1.87	556	−61.44
0.30	435	1.87	621	−56.93
0.40	435	1.87	398	−72.40
0.50	435	1.87	237	−83.56
0.60	435	1.87	229	−84.12
0.70	435	1.87	283	−80.37
0.80	496	16.16	224	−84.47
0.90	496	16.16	226	−84.33
1.00	496	16.16	236	−83.63
1.30	547	28.10	142	−90.15
1.60	547	28.10	107	−92.58
2.00	574	34.43	76	−94.73

Source: Adapted from Huang *et al.* (2012*a*). Reproduced with permission of Springer.

9.5 Randomly Generated Examples

In this section, we generate a set of 50 random nets for manufacturing systems to further test A*, DWA, and IDWA with the same heuristic function h_H. These random problems, which are generated by randomly selecting and linking predefined Petri net modules (Mejía, 2002), have the following characteristics. Each manufacturing system has three kinds of shared resources and four different jobs. Each job has three tasks and 40% of the tasks have two alternative routes and 50% of the alternative routes have one operation while the rest have two sequential operations. Between any two sequential operations, an intermediate buffer exists to store the parts that are ready for the next operation. Each operation is assigned a random integer operation time from a uniform distribution [50, 100]. The lot size of each job is limited to one.

Note that the DWA algorithm needs to analyze the structure of these nets and predict an upper bound of the depths of the goal nodes in the nets. For those random nets, since an alternative route of a task has at most four places (including two activity places and two places for intermediate buffers) and each job has

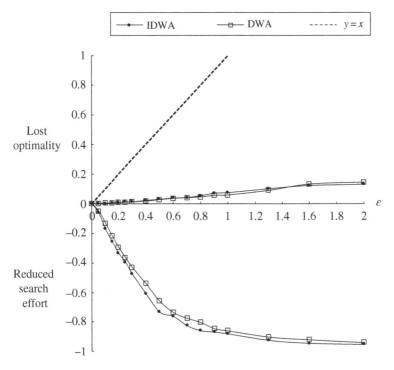

Figure 9.2 Performances of IDWA and DWA, averaged over 50 random SC-nets. *Source:* Adapted from Huang *et al.* (2012a). Reproduced with permission of Springer.

three tasks, the maximal depth of a job is $4 \times 3 = 12$. In addition, there are four jobs in each net. Thus, an upper bound of the depths of solutions of those nets is $d(S_G) = 12 \times 4 = 48$.

Figure 9.2 shows the mean percentage of lost optimality ($\mathcal{R}_c$) and the mean percentage of reduced search effort ($\mathcal{R}_E$) of IDWA and DWA when compared with the A* algorithm by using different ε-values. The line $y = x$ is included to show the theoretical limit of $\mathcal{R}_c$. We can see that it is possible for IDWA and DWA to return a solution far better than the suggested limit. The following merits of IDWA can also be concluded. First, the dependency on an anticipated depth has been avoided by IDWA and the qualities of solutions ($\mathcal{R}_c$) of IDWA are not worse than those of DWA. Second, the mean percentages of reduced search effort ($\mathcal{R}_E$) of IDWA are lower than those of DWA. This phenomenon is to be expected since IDWA is target-oriented, while DWA is source-oriented. The factor $1 - d(S)/d(S_G)$ of DWA decreases as the search moves away from the initial state, whereas $h(S)/h(S_0)$ of IDWA decreases as the search approaches the goal state. This means that $1 - d(S)/d(S_G)$ of DWA repels the search away from the initial state, whereas $h(S)/h(S_0)$ of IDWA attracts the search towards the goal state. The subtle difference is that if a search moves away from both the initial state and the goal state at the same time, then IDWA tends to correctly penalize it, but DWA rewards it.

9.6 Another Controllable Heuristic Search

This section introduces another controllable heuristic search strategy that combines A* search and depth-first search within the reachability graphs of place-timed SC-nets to schedule RASs. First, we present them within the reachability graph of an SC-net.

9.6.1 A* Search and Depth-First Search

To find a schedule path for a Petri net, the search space is the reachability graph of the net, which contains all reachable markings and the edges connecting these markings. At a reachable marking, firing an enabled transition changes the tokens' distribution of the net and generates a new marking. A sequence of transition firings generates a sequence of markings. Thus, the search problem in the reachability graph becomes finding a firing sequence of the transitions in the graph from an initial marking to a goal marking. In Section 8, time information is introduced into some places of SC-nets for the modeling and scheduling of RASs and an A* heuristic search algorithm is developed for place-timed SC-net execution. The most important aspects of the algorithm are the consideration of both markings and tokens' remaining times in model evolution and avoidance of generating the entire reachability graph to find a schedule. The A* search for SC-nets is given in Algorithm 9.2.

Algorithm 9.2
```
A* algorithm for SC-nets:
```
    1) Place the initial state S_0 on the list OPEN.
    2) If OPEN is empty, terminate with failure.
    3) Remove the first state $S = (M, R)$ from OPEN and put S on the list CLOSED.
    4) If S equals the goal state S_G, construct the path from S back to S_0 and terminate.
    5) Find the set of transitions enabled at M.
    6) For each transition t enabled at M, decrease the tokens' remaining times to enable t with respect to both M and R, generate an immediate succeeding state $S' = (M', R')$ by firing t, set a pointer from S' to S, and calculate $g(S') = g(S) + c(S, S')$, $h(S')$, and $f(S') = g(S') + h(S')$.
        a) If there exists a state $S^O = (M^O, R^O)$ on OPEN such that $M' = M^O$, $R' = R^O$, and $g(S') < g(S^O)$, delete S^O from OPEN and insert S' on OPEN.

      b) If there exists a state $S^C = (M^C, R^C)$ on CLOSED such that $M' = M^C$ and $R' = R^C$, and $g(S') < g(S^C)$, delete S^C from CLOSED and insert S' on OPEN.

      c) If both lists do not contain a state with M' and R', insert S' on OPEN.

    7) Reorder OPEN in the non-decreasing magnitude of f-values.

    8) Go to Step 2).

In an SC-net, a state is denoted as $S = (M, R)$ where M represents the tokens' distribution and R denotes a tokens' remaining time matrix. The evaluation function $f(S)$ of a state S is calculated by $f(S) = g(S) + h(S)$ in which $g(S)$ represents the cost or makespan of the partial schedule determined so far and the heuristic function $h(S)$ represents an estimate of the remaining cost (makespan) to reach a goal state S_G. Note that S_G can be the state in which all part tokens are moved into their end places in many SC-nets.

At each iteration of the A* search algorithm, a most promising state which has the minimal f-value in the list OPEN is chosen for expansion. If the chosen state equals the goal state, the search terminates; otherwise, the state is expanded by firing all of its enabled transitions to generate immediate succeeding states. The new states are picked and added to the list OPEN that contains states that have been generated but not yet expanded. Again one of the most promising states is selected for expansion and the process continues.

If the heuristic function h is a lower bound of all complete solutions descending from the current state, i.e.,

$$h(S) \leq h^*(S), \ \forall S \tag{9.7}$$

where $h^*(S)$ is the lowest cost of paths going from S to the goal state, then h is said to be admissible. If an admissible heuristic function is used by the A* algorithm, an optimal schedule can be obtained if such a schedule exists. But for sizable systems, the A* algorithm is often difficult to terminate in a reasonable amount of time. One of the main reasons is that in many problems, A* spends a large amount of time discriminating among paths whose costs do not vary significantly from each other, which forces A* to spend a long time selecting the best among roughly equal candidates and prevents A* from terminating with a suboptimal but acceptable solution. In the literature, some improved methods (Baruwa *et al.*, 2015; Lee & Lee, 2010; Luo *et al.*, 2015; Mejía, 2002; Moro *et al.*, 2002b; Xiong & Zhou, 1998; Yu *et al.*, 2003) have been proposed to reduce the search effort of the A* algorithm, but the quality of the obtained solutions cannot be guaranteed. This section presents a heuristic scheduling approach by combining the A* algorithm with the depth-first strategy based on the execution of place-timed SC-nets. This approach endows A* with a stronger depth-first component to search faster but also guarantees that the solution obtained at termination is not worse than the optimal one by a factor greater than a given value. In the following, we first introduce the depth-first (DF) search algorithm for SC-net scheduling as shown in Algorithm 9.3.

Algorithm 9.3

```
Depth-first Algorithm:
    1) Place the initial state S₀ on the list OPEN.
    2) If OPEN is empty, terminate with failure.
    3) Remove the first state S = (M, R) from OPEN and put S on the
list CLOSED.
    4) If S equals the goal state S_G, construct the path from S back
to S₀ and terminate.
    5) Find the set of transitions enabled at M.
    6) For each transition t enabled at M, decrease the tokens'
remaining times to enable t, generate an immediate succeeding state
S' = (M', R') by firing t, set a pointer from S' to S, and calcu-
late g(S') = g(S) + c(S, S').
        a) If there exists a state S^O = (M^O, R^O) on OPEN such that
M' = M^O, R' = R^O, and g(S') < g(S^O), delete S^O from OPEN and insert S' in
the top of OPEN.
        b) If there exists a state S^C = (M^C, R^C) on CLOSED such that
M' = M^C and R' = R^C, and g(S') < g(S^C), delete S^C from CLOSED and insert
S' in the top of OPEN.
        c) If both lists do not contain a state with M' and R',
insert S' in the top of OPEN.
    7) Go to Step 2).
```

Note that the A* algorithm expands a state with the minimal f-value in OPEN at each iteration, while the depth-first algorithm gives priority to a state at the deepest level in OPEN at each iteration. It guarantees that no state at a depth d is selected for expansion as long as a state whose depth is greater than d exists in OPEN. In the following, several cases with different lot sizes of an example SC-net are tested with the A* algorithm and the depth-first algorithm.

From Chapter 8, we see that RASs can be modeled by place-timed SC-nets via a top-down method or a bottom-up method as in Section 8.2.3 or a conversion method that converts an existing untimed net into an SC-net as in Section 8.2.2. In the sequel, the bottom-up synthesis method is adopted to model a flexible manu-facturing system with SC-nets. The obtained model is used as a running example in this section. The manufacturing system consists of three types of shared resources R_1–R_3 and four types of jobs P_1–P_4. The job requirements are given in Table 9.3 where each job has three operations and the processing time of each operation is shown in parentheses. Suppose that between any two sequential operations of the system, an intermediate buffer exists to store parts for the next operation. The bottom-up modeling process is briefed as follows. First, model a Petri net for each job based on the production sequence and the use of resources for each opera-tion. Next, merge these models to obtain a complete Petri net through the shared resource places. Then, the lot sizes of jobs are represented by the tokens in the

Table 9.3 Job requirements of the example system.

	P_1	P_2	P_3	P_4
Operation$_1$	$R_1(2)$	$R_3(4)$	$R_1(3)$	$R_2(3)$
Operation$_2$	$R_2(3)$	$R_1(2)$	$R_3(5)$	$R_3(4)$
Operation$_3$	$R_3(4)$	$R_2(2)$	$R_2(3)$	$R_1(3)$

Source: Adapted from Huang *et al.* (2008). Reproduced with permission of Taylor & Francis.

corresponding start places and the processing time of each operation is added onto the corresponding activity place. Figure 9.3 shows the obtained SC-net of the manufacturing system. Note that activity places without an integer number refer to the intermediate buffers.

The DF algorithm and A* algorithm with the admissible heuristic function h_H introduced in Section 9.3 are used for SC-net scheduling. In path finding, i) if more than one state has the minimal f-value in OPEN, the latest introduced state is first selected for expansion, i.e., the tie-breaking rule is last-in-first-out; ii) if more than one enabled transition exists at a state, these transitions fire in order of their subscripts. The scheduling results in terms of makespan c and the number of expanded states N_E are shown in Table 9.4. From the table, we see that Algorithm 9.2 (A*) finds the optimal solutions at the expense of computational complexity, while Algorithm 9.3 (DF) reduces the computational complexity at the expense of optimality. For many practical scheduling problems, obtaining a suboptimal solution in less time may be desired. Thus, a combination of A* search and DF search is a promising way to schedule such RASs.

9.6.2 Controllable Hybrid Heuristic Search

Algorithm 9.4 gives an A*-DF search algorithm that combines A* algorithm with depth-first search to improve the search efficiency by relaxing the admissibility condition.

Algorithm 9.4

```
Algorithm A*-DF
    1) Set a desired value of ε such that ε ≥ 0.
    2) Place the initial state S₀ on the list OPEN.
    3) If OPEN is empty, terminate with failure.
    4) Select the states in OPEN whose f-values do not devi-
ate from the f-value of the first state in OPEN by a factor greater
than 1+ε to constitute the list SELECT, i.e.,
```

$$\text{SELECT} = \{S | f(S) \leqslant (1 + \varepsilon) \cdot \min\nolimits_{S' \in \text{ OPEN}} f(S')\}.$$

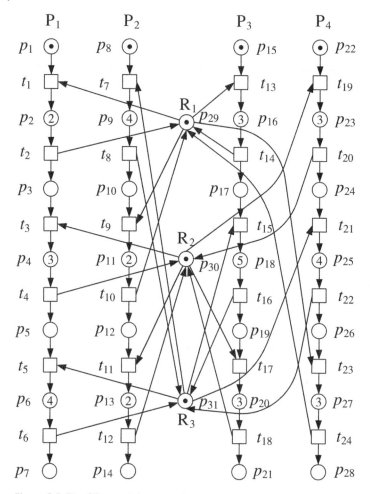

Figure 9.3 The SC-net of the example system. Source: Adapted from Huang *et al.* (2008). Reproduced with permission of Taylor & Francis.

4) Find the deepest state $S = (M, R)$ from SELECT.

5) If S equals the goal state S_G, construct the path from S back to S_0 and terminate.

6) Find the set of transitions enabled at M.

7) For each transition t enabled at M, decrease the tokens' remaining times to enable t with respect to both M and R, generate an immediate succeeding state $S' = (M', R')$ by firing t, set a pointer from S' to S, and calculate $g(S') = g(S) + c(S, S')$, $h(S')$, and $f(S') = g(S') + h(S')$.

      a) If there exists a state $S^O = (M^O, R^O)$ on OPEN such that $M' = M^O$, $R' = R^O$, and $g(S') < g(S^O)$, delete S^O from OPEN and insert S' on OPEN.

Table 9.4 Scheduling results of the example system.

Lot sizes				c		N_E	
P_1	P_2	P_3	P_4	A*	DF	A*	DF
1	1	1	1	17	38	41	25
2	2	2	2	34	76	460	49
3	3	3	3	51	114	1113	73
5	5	5	5	85	190	1765	121
10	10	10	10	170	380	3941	241

Source: Adapted from Huang *et al.* (2008). Reproduced with permission of Taylor & Francis.

      b) If there exists a state $S^C = (M^C, R^C)$ on CLOSED such that $M' = M^C$ and $R' = R^C$, and $g(S') < g(S^C)$, delete S^C from CLOSED and insert S' on OPEN.
      c) If both lists do not contain a state with M' and R', insert S' on OPEN.
    8) Reorder OPEN in the non-decreasing order of f-values.
    9) Go to Step 3).

Algorithm 9.4 uses three lists: OPEN, SELECT, and CLOSED. SELECT is a sublist of OPEN and it only contains states whose f-values do not deviate from the lowest f-value in OPEN by a factor greater than $1 + \varepsilon$. At each iteration of the algorithm, the deepest state in SELECT is picked for expansion. The reason for this operation is simple. All states in SELECT represent roughly equal solution paths in terms of the f-values. Therefore, rather than spending time deciding which state among them is the best for expansion, it makes more sense to directly select the deepest one from them to accelerate the search process.

For sizable scheduling problems, if we cannot afford the computational time required by A* search, we can employ the A*-DF combination to cut down the computational time. It is at the expense of narrowing down evaluation scope and risks missing an optimal schedule. However, the quality of the obtained schedule is controllable, that is, the A*-DF algorithm can find a solution whose cost does not exceed the optimal one by more than a factor of $1 + \varepsilon$.

Theorem 9.3 *The A*-DF algorithm can find a solution whose cost does not exceed the optimal one by more than a factor $1 + \varepsilon$.*

Proof: *Adopt the following notations:*
    S_1: *a state that has the smallest f-value in OPEN;*
    S_t: *a termination state that is chosen for expansion and found equal to the goal state S_G;*

S_s: *the shallowest OPEN state on an optimal path;*
$c(S_t) = f(S_t) = g(S_t)$: *the cost of a found path;*
$c^*(S_t)$: *the lowest cost among all paths from the initial state S_0 to S_G;*
We have:
$f(S_s) \leq c^*$ *since h is admissible*
$f(S_1) \leq f(S_s)$ *because OPEN is f-ordered*
$f(S_t) \leq f(S_1)(1 + \varepsilon)$ *since S_t is chosen from SELECT*
Therefore:
$c(S_t) = f(S_t) \leq f(S_s)(1 + \varepsilon) \leq c^*(1 + \varepsilon).$ ∎

To demonstrate it, three cases of the SC-net in Figure 9.3 with different lot sizes $(3, 3, 3, 3)$, $(5, 5, 5, 5)$, and $(10, 10, 10, 10)$ are tested by using the A*-DF algorithm with the admissible heuristic function h_H and different ε-values. The scheduling results in terms of makespan c, the percentage of lost optimality $\mathcal{R}_c$, the number of expanded states N_E, the percentage of reduced search effort $\mathcal{R}_E$, the computational time $\mathcal{T}$, and the percentage of the reduced computational time $\mathcal{R}_T$ are shown in Tables 9.5–9.7 for lot sizes $(3, 3, 3, 3)$, $(5, 5, 5, 5)$, and $(10, 10, 10, 10)$, respectively. In these tables, the percentage of lost optimality $\mathcal{R}_c$, which is the comparison

Table 9.5 Scheduling results for the lot size (3, 3, 3, 3) by using A*-DF.

ε	c	$\mathcal{R}_c$(%)	N_E	$\mathcal{R}_E$(%)	$\mathcal{T}$ (s)	$\mathcal{R}_T$(%)
0.00	51	0.00	1126	0.00	5.81	0.00
0.01	51	0.00	1126	0.00	5.75	−1.03
0.02	52	1.96	488	−56.66	2.81	−51.64
0.03	52	1.96	488	−56.66	2.35	−59.55
0.04	53	3.92	690	−38.72	3.62	−37.69
0.05	53	3.92	690	−38.72	3.38	−41.82
0.06	54	5.88	309	−72.56	1.78	−69.36
0.07	54	5.88	309	−72.56	1.78	−69.36
0.08	55	7.84	766	−31.97	3.96	−31.84
0.09	55	7.84	766	−31.97	3.68	−36.66
0.10	56	9.80	370	−67.14	1.84	−68.33
0.11	56	9.80	370	−67.14	2.41	−58.52
0.12	57	11.76	275	−75.58	1.75	−69.88
0.13	57	11.76	275	−75.58	1.37	−76.42
0.14	58	13.73	232	−79.40	1.17	−79.86
0.15	58	13.73	232	−79.40	1.18	−79.69

Source: Adapted from Huang *et al.* (2008). Reproduced with permission of Taylor & Francis.

Table 9.6 Scheduling results for the lot size (5, 5, 5, 5) by using A*-DF.

ε	c	$\mathcal{R}_c$(%)	N_E	$\mathcal{R}_E$(%)	$\mathcal{T}$(s)	$\mathcal{R}_T$(%)
0.00	85	0.00	1765	0.00	10.69	0.00
0.01	85	0.00	1765	0.00	9.49	−11.23
0.02	86	1.18	1572	−10.93	8.52	−20.30
0.03	87	2.35	800	−54.67	5.26	−50.80
0.04	88	3.53	1213	−31.72	8.15	−23.76
0.05	89	4.71	1086	−38.47	5.41	−49.39
0.06	90	5.88	872	−50.59	5.58	−47.80
0.07	90	5.88	872	−50.59	5.13	−52.01
0.08	91	7.06	695	−60.62	4.01	−62.49
0.09	92	8.24	724	−58.98	4.05	−62.11
0.10	93	9.41	388	−78.02	1.88	−82.41
0.11	94	10.59	205	−88.39	1.19	−88.87
0.12	95	11.76	172	−90.25	1.05	−90.18
0.13	96	12.94	141	−92.01	0.67	−93.73
0.14	96	12.94	141	−92.01	0.71	−93.36
0.15	96	12.94	141	−92.01	0.96	−91.02

Source: Adapted from Huang *et al.* (2008). Reproduced with permission of Taylor & Francis.

of the makespans (costs) needed by the schedules of different algorithms, is defined as

$$\mathcal{R}_c = \frac{c(\text{A*-DF}) - c(\text{A*-DF}_{\varepsilon=0})}{c(\text{A*-DF}_{\varepsilon=0})} \times 100\%, \tag{9.8}$$

the percentage of reduced search effort $\mathcal{R}_E$, which is defined in Huang et al. (2008) as the comparison of the numbers of states expanded by different algorithms, is

$$\mathcal{R}_E = \frac{N_E(\text{A*-DF}) - N_E(\text{A*-DF}_{\varepsilon=0})}{N_E(\text{A*-DF}_{\varepsilon=0})} \times 100\%, \tag{9.9}$$

and the percentage of reduced computational time $\mathcal{R}_T$ is defined as

$$\mathcal{R}_T = \frac{\mathcal{T}(\text{A*-DF}) - \mathcal{T}(\text{A*-DF}_{\varepsilon=0})}{\mathcal{T}(\text{A*-DF}_{\varepsilon=0})} \times 100\%. \tag{9.10}$$

Note that the A*-DF algorithm with $\varepsilon = 0$ is equal to the A* algorithm except that if more than one state in OPEN has the minimal f-value, the A*-DF algorithm with $\varepsilon = 0$ chooses the deepest one for expansion while the A* algorithm chooses the newest one for expansion since the used tie-breaking rule is last-in-first-out.

From the results shown in these tables, we can see that the presented A*-DF algorithm with $\varepsilon > 0$ yields a sharp drop in the search effort and computational

Table 9.7 Scheduling results for the lot size (10, 10, 10, 10) by using A*-DF.

ε	c	$R_c(\%)$	N_E	$R_E(\%)$	$T(s)$	$R_T(\%)$
0.00	170	0.00	3941	0.00	32.24	0.0
0.01	171	0.59	1583	−59.83	12.43	−61.45
0.02	173	1.76	1535	−61.05	11.81	−63.37
0.03	175	2.94	1618	−58.94	11.42	−64.58
0.04	176	3.53	2442	−38.04	12.01	−62.75
0.05	178	4.71	2426	−38.44	11.89	−63.12
0.06	180	5.88	2447	−37.91	14.28	−55.71
0.07	181	6.47	2242	−43.11	10.58	−67.18
0.08	183	7.65	2067	−47.55	22.91	−28.94
0.09	185	8.82	1863	−52.73	10.95	−66.04
0.10	187	10.00	1808	−54.12	8.41	−73.91
0.11	188	10.59	1604	−59.30	7.93	−75.40
0.12	190	11.76	814	−79.35	3.65	−88.68
0.13	192	12.94	446	−88.68	2.44	−92.43
0.14	193	13.53	413	−89.52	1.83	−94.32
0.15	195	14.71	333	−91.55	1.51	−95.32

Source: Adapted from Huang *et al.* (2008). Reproduced with permission of Taylor & Francis.

time since R_E and R_T are very low, while the solution found at termination is not worse than the optimal one by a factor greater than $1 + \varepsilon$, i.e., $R_c \leq \varepsilon$.

9.7 Illustrative Results

In this section, we test the A*-DF algorithm with a more complex SC-net for a flexible manufacturing system that is adapted from Chen and Li (2011), Ezpeleta *et al.* (1995), and Huang *et al.* (2015c). The manufacturing system has three robots R_1–R_3, four kinds of machines M_1–M_4, three loading buffers I_1–I_3, and three unloading buffers O_1–O_3. Three types of parts, P_1–P_3, are processed in the system. The production sequences are:

$$P_1: I_1 \rightarrow R_1 \rightarrow M_1 \rightarrow R_2 \rightarrow M_2 \rightarrow R_3 \rightarrow O_1$$
$$\text{or } I_1 \rightarrow R_1 \rightarrow M_3 \rightarrow R_2 \rightarrow M_4 \rightarrow R_3 \rightarrow O_1$$
$$P_2: I_2 \rightarrow R_2 \rightarrow M_2 \rightarrow R_2 \rightarrow O_2$$
$$P_3: I_3 \rightarrow R_3 \rightarrow M_4 \rightarrow R_2 \rightarrow M_3 \rightarrow R_1 \rightarrow O_3$$

Table 9.8 Operation times of activity places.

Place	Operation time	Place	Operation time
p_2	2	p_{11}	1
p_3	4	p_{12}	3
p_4	5	p_{13}	4
p_6	3	p_{15}	6
p_7	2	p_{16}	3
p_8	6	p_{17}	4
p_9	2	p_{18}	6
p_{10}	4	p_{19}	2

Each part has three units to be processed. The processing time of each operation is shown in Table 9.8. Its place-timed SC-net is given in Figure 9.4 where each idle place of the original net has been split into a start place and an end place to discriminate a goal state from the other states for scheduling. The net has 20 transitions and 29 places in which $P_S = \{p_1, p_5, p_{14}\}$, $P_E = \{p_{27}, p_{28}, p_{29}\}$, $P_R = \{p_{20} - p_{26}\}$, and the rest belong to the activity places P_A. The scheduling results are summarized in Table 9.9.

From the results, we can see that the A* algorithm obtains a schedule that has the smallest cost and is guaranteed to be optimal since an admissible heuristic function h_H is used, but the A* algorithm is time-consuming and needs to expand a lot of states. For example, the computational time of A* is about 330 seconds and its number of expanded states is 36 809. On the other hand, the DF algorithm can obtain a schedule in a much shorter time but the quality of the schedule is unsatisfactory (the cost or makespan of the schedule is 62). However, the presented A*-DF algorithm with ε expands much fewer states than the A* algorithm and the obtained solution's cost does not exceed the optimal cost by a factor greater than $1 + \varepsilon(\varepsilon \geq 0)$. In addition, the following phenomena can be observed. i) The number of expanded states is not always a monotonically decreasing function of ε. This phenomenon is to be expected. This is because while A* guarantees that no state with $f(S) > c^*$ (where c^* stands for the optimal cost) would ever be expanded, A*-DF only guarantees that the states satisfying $f(S) > c^*(1 + \varepsilon)$ are excluded from expansion. Consequently, some states satisfying the condition $c^*(1 + \varepsilon) > f(S) > c^*$ would be expanded by A*-DF and not by A*. The number of such states could, in some cases, exceed the savings provided by A*-DF. ii) Although A*-DF is designed to tolerate the worst increase of cost up to εc^*, the actual increase of cost is substantially lower. For example, in A*-DF with $\varepsilon = 0.07$, the percentage of lost optimality is $\mathcal{R}_c = 2.33\%$, which is much smaller 7% implied by $\varepsilon = 0.07$.

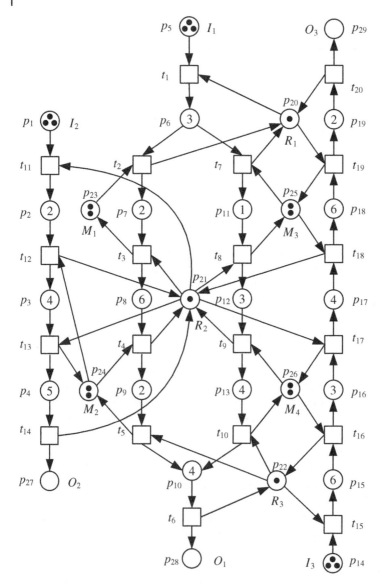

Figure 9.4 SC-net of a more complex system.

9.8 Concluding Remarks

This chapter presents an admissible heuristic function and two controllable scheduling approaches for the scheduling of place-timed SC-nets that can naturally model RASs with concurrent activities, shared resources, precedence

Table 9.9 Scheduling results of the SC-net in Figure 9.4 (A*: $c = 43$, $N_E = 36\,809$, $\mathcal{T} = 330$ seconds; DF: $c = 62$, $N_E = 302$, $\mathcal{T} = 1.62$ seconds).

ε	c	$\mathcal{R}_c(\%)$	N_E	$\mathcal{R}_E(\%)$	$\mathcal{T}(s)$	$\mathcal{R}_T(\%)$
0.00	43	0	12 026	0	82.40	0.00
0.01	43	0	12 026	0	75.02	−8.96
0.02	43	0	12 026	0	80.57	−2.22
0.03	43	0	10 778	−10.38	56.48	−31.46
0.04	43	0	10 778	−10.38	52.89	−35.81
0.05	44	2.33	1 845	−84.66	13.08	−84.13
0.06	44	2.33	1 845	−84.66	12.71	−84.58
0.07	44	2.33	1 845	−84.66	11.63	−85.89
0.08	45	4.65	4 087	−66.02	27.66	−66.43
0.09	45	4.65	4 087	−66.02	27.43	−66.71
0.10	46	6.98	2 539	−78.79	16.54	−79.93

constraints, and operation times. For the first approach, the number of expanded states is dramatically reduced when compared with the A* search and the quality of the search result is controllable when an admissible heuristic function is used. In addition, the method does not need to estimate the depth of final solutions in advance as the DWA algorithm and the performances are better than DWA in terms of the number of expanded states and the result's quality. The second approach combines the A* algorithm and the depth-first strategy to accelerate the scheduling process for such models. It can invoke quicker termination conditions and the quality of the obtained solution is also controllable, i.e., the cost of the solution does not exceed the optimal cost by a factor greater than $1 + \varepsilon$ ($\varepsilon \geq 0$).

9.9 Bibliographical Notes

For the scheduling of place-timed SC-nets of RASs, this chapter presents two controllable heuristic A* methods that can speed up the search process while the result's quality is controllable. The material of this chapter mainly comes from Huang *et al.* (2008, 2012a). For classifications and applications of timed Petri nets, please refer to Wang (1998). The modeling and job scheduling with Petri nets and other tools are in the book by Hrúz and Zhou (2007). Heuristic search strategies can be seen in Edelkamp and Schroedl (2012) and Pearl (1984). More search approaches with artificial intelligence can be seen in Russell and Norvig (2010).

10

Hybrid Heuristic Search

This chapter presents a hybrid scheduling strategy that combines two search schemes to schedule resource allocation systems (RASs) based on an SC-net framework. RASs may have different lot sizes, strict precedence constraints, multiple copies of resources, and concurrent activities. To reduce the computational complexity in RAS scheduling, this chapter presents an elaborate search scheme that performs an A* algorithm locally and a backtracking search globally in the reachability graphs of SC-nets. Some numerical experiments are carried out to show that the method is more efficient than other similar methods that also combine the A* algorithm with the backtracking strategy.

10.1 Introduction

In resource allocation systems (RASs), system scheduling is a typical combinatorial optimization problem that determines start time and allocations of different shared resources to operations. A good scheduling method for these systems should include two characteristics: easy formulation of RASs and quick identification of acceptable solutions with small computational effort. Many industrial and academic communities are focusing on developing such methods to schedule real-world RASs.

A problem observed in RAS scheduling is the difficulty in finding an optimal or near-optimal solution in a reasonable amount of time for a sizable system. In the literature, some efficient methods have been proposed to reduce such search effort. For example, Lee and DiCesare (1994) propose an A*-based search method within reachability graphs to schedule flexible manufacturing systems as a typical kind of RASs. The method only needs to explore a partial reachability graph to obtain a result. In addition, it can guarantee the optimality of the result if an admissible heuristic function is used. However, such a method suffers from the state

Supervisory Control and Scheduling of Resource Allocation Systems: Reachability Graph Perspective,
First Edition. Bo Huang and MengChu Zhou.
© 2020 The Institute of Electrical and Electronics Engineers, Inc. Published 2020 by John Wiley & Sons, Inc.

explosion problem, i.e., the number of considered states increases exponentially with the net size. So, it can only be applied to small-scale systems. To alleviate this problem, Xiong and Zhou (1998) propose a hybrid strategy that first uses an A* search and then a backtracking search (denoted as a BF-BT algorithm), or vice versa (denoted as a BT-BF algorithm). Both BF-BT and BT-BF can obtain a suboptimal schedule for the underlying system with less computational effort.

This chapter presents another hybrid scheduling method based on SC-nets of RASs. The search scheme adopts a more elaborate and efficient scheme that performs an A* search locally and a backtracking search globally. Several cases of an RAS with different lot sizes are tested with the presented algorithm, BF-BT algorithm, and BT-BF algorithm. These methods are also tested and compared with a set of randomly generated and more complex RASs that have limited buffers, alternative routes, and operations that need multiple resources.

10.2 A*-BT Combinations

We already know that the results obtained by an A* algorithm are optimal if an admissible heuristic function is used by the algorithm. But for sizable problems, the A* algorithm seldom finds a solution in a reasonable amount of time. To reduce the memory space and computational time required by the A* search within reachability graphs, Xiong and Zhou (1998) propose two hybrid search methods (BF-BT and BT-BF) that combine an A* search with a backtracking strategy within the reachability graph of a Petri net. The BF-BT algorithm is depicted in Figure 10.1a where the A* strategy is applied at the top of the search graph (represented by a shaded area with an irregular frontier) and a backtracking strategy at the bottom (represented by a left-to-right arrow). As soon as a given depth d_{max} is reached, the backtracking search takes over from the best node found by the A* algorithm until the entire graph beneath that node is traversed. Another algorithm BT-BF

(a) 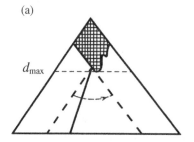 (b)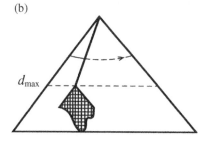

Figure 10.1 (a) The BF-BT algorithm; (b) The BT-BF algorithm. Source: Adapted from Huang *et al.* (2010). Reproduced with permission of Springer.

Figure 10.2 The algorithm performing A* locally and BT globally. Source: Adapted from Huang *et al.* (2010). Reproduced with permission of Springer.

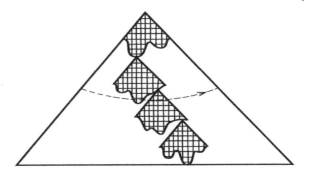

is shown in Figure 10.1b where the backtracking search is employed at the top of the graph and A* is used at the bottom. The backtracking search is applied until a given depth d_{max} is reached. At this point, the A* search is started from the node of the backtracking search until it returns a solution or exists with a failure.

Since the performance of an A* algorithm is usually at its best at the bottom of the search graph (Xiong & Zhou, 1998), BT-BF performs much better than BF-BT. However, the BT-BF method has two problems: i) Some important decisions with respect to the quality of a result may happen at the early stages of the search process. So, BT-BF increases the likelihood of missing some critical candidates since it employs a backtracking search instead of a best-first search at the early stages; and ii) The depth of the solution is usually unknown in advance, so it is hard to decide when and where to trigger the transition from BT to BF is appropriate.

This chapter presents a more elaborate scheme that performs A* locally and BT globally as depicted in Figure 10.2. It is an iterative search process. At each iteration, it begins in an A* manner and checks the number of states in the OPEN list after expanding a state to generate its immediate successors. Once the number of states in OPEN exceeds a threshold N_{max}, it regards all the states in OPEN as the children of the root state at this iteration and submits them to a backtracking search. The backtracking search selects the best state among these states with an evaluation function f and moves the state to the next iteration as a root state that starts the A* search until the number of states in a new OPEN list exceeds N_{max} again or the new OPEN list becomes empty. If the number of states in the new OPEN exceeds N_{max}, a new iteration continues like this; if the new OPEN becomes empty, then another best state from the OPEN list at the last iteration is picked as a new root state of the A* search at this iteration. The detailed procedures of the hybrid search are given in Algorithm 10.1.

Algorithm 10.1

```
The algorithm performing A* locally and BT globally for SC-nets:
    1)  i = 0.
    2)  Place the initial state S_0 on the list OPEN_i.
```

3) If $i > 0$, remove the first state $S = (M, R)$ from list OPEN_{i-1} and put S on list OPEN_i.

4) If OPEN_i is empty, then check i: if $i = 0$, terminate with failure; otherwise, go to Step 3).

5) Remove the first state S from OPEN_i and put S on the list CLOSED.

6) If S equals the goal state S_G, construct the path from S back to S_0 and terminate.

7) Find the set of transitions enabled at M.

8) For each transition t enabled at M, decrease the tokens' remaining times to enable t at both M and the tokens' remaining times R, generate an immediate succeeding state $S' = (M', R')$ by firing t, set a pointer from S' to S, and calculate $g(S') = g(S) + c(S, S')$, $h(S')$, and $f(S') = g(S') + h(S')$.

    a) If there exists a state $S^O = (M^O, R^O)$ on OPEN such that $M' = M^O$, $R' = R^O$, and $g(S') < g(S^O)$, delete S^O from OPEN_i and insert S' on OPEN_i.

    b) If there exists a state $S^C = (M^C, R^C)$ on CLOSED such that $M' = M^C$ and $R' = R^C$, and $g(S') < g(S^C)$, delete S^C from CLOSED and insert S' on OPEN_i.

    c) If both lists do not contain a state with M' and R', insert S' on OPEN_i.

9) Reorder OPEN_i in the non-decreasing magnitude of f-values.

10) If the number of states in OPEN_i exceeds $N_{\max}$, then $i = i + 1$ and go to Step 3); otherwise, go to Step 4).

In summary, Algorithm 10.1 is an informed depth-first search where a memory-limited A* search is applied to expand each state of the depth-first search. All the states generated by the A* search are considered as immediate successors of the state in the depth-first search. As an example, we test the SC-net in Figure 9.3 with three sets of lot sizes (5, 5, 2, 2), (8, 8, 4, 4), and (10, 10, 6, 6) by using different search algorithms with the same heuristic function h_H given in Section 9.3. The scheduling results of Algorithm 10.1 in terms of the makespan c, the number of expanded states N_E, and the computational time $\mathcal{T}$ are shown in Tables 10.1–10.3 for the cases with different lot sizes (5, 5, 2, 2), (8, 8, 4, 4), and (10, 10, 6, 6). The corresponding results of the A* search, backtracking method, BF-BT algorithm, BT-BF algorithm, and presented method are shown in these tables for comparisons.

Note that BF-BT, BT-BF, and Algorithm 10.1 can cut down the computational complexity in the pathfinding by narrowing the evaluation scope. However, this is at the expense of losing the optimality of the results compared to a pure A* search. So, we use $\mathcal{R}_E$ to denote the percentage of the change in computational complexity, comparing the number of expanded states of a hybrid method with that of the A* algorithm, as follows:

$$\mathcal{R}_E = \frac{N_E(\text{hybrid}) - N_E(\text{A}^*)}{N_E(\text{A}^*)} \times 100\%. \tag{10.1}$$

Table 10.1 Scheduling results of the SC-net in Figure 9.3 with the lot size (5, 5, 2, 2).

Algorithm	$d_{max}(N_{max})$	c	N_E	$T(s)$
A*	—	58	3437	14.00
BT	—	105	85	571.00
	20	94	571	0.65
	40	85	1607	4.00
BF-BT	50	79	2123	6.00
	60	74	2775	8.00
	80	64	3308	11.00
	20	88	248	0.38
	40	80	484	0.80
BT-BF	50	70	1247	3.60
	60	64	1520	6.50
	80	62	1687	7.00
	1	75	230	0.17
	5	68	358	0.31
	10	67	634	0.62
Algorithm 10.1	15	65	540	0.46
	20	62	946	0.78
	25	62	921	0.93
	30	60	1423	1.12

Source: Adapted from Huang *et al.* (2010). Reproduced with permission of Springer.

In addition, we use R_c to denote the percentage of lost optimality, which is the comparison of the makespan of a schedule obtained by a hybrid method with that of the A* search, as follows:

$$R_c = \frac{c(\text{hybrid}) - c(\text{A}^*)}{c(\text{A}^*)} \times 100\%. \tag{10.2}$$

The relations between R_E to R_c of different methods for the three cases with different lot sizes (5, 5, 2, 2), (8, 8, 4, 4), and (10, 10, 6, 6) are shown in Figures 10.3–10.5, respectively.

From the results, the following phenomena can be observed.

1) Algorithm 10.1 that employs A* locally and BT globally within reachability graphs performs better than both BF-BT and BT-BF algorithms owing to the below reasons. First, the performance of the A* search is at its best when the

Table 10.2 Scheduling results of the SC-net in Figure 9.3 with the lot size (8, 8, 4, 4).

Algorithm	$d_{max}(N_{max})$	c	N_E	$T(s)$
A*	—	100	9438	112.00
BT	—	198	145	0.23
	40	168	3888	24.00
	60	154	5234	38.00
BF-BT	80	140	7699	49.00
	100	127	8819	90.00
	120	108	9233	104.00
	40	163	585	1.40
	60	140	1590	7.00
BT-BF	80	121	2873	18.00
	100	112	4545	36.00
	120	104	8045	76.00
	1	128	422	0.31
	5	116	570	0.46
	10	111	782	0.62
Algorithm 10.1	15	112	1047	1.56
	20	109	1694	2.65
	25	105	1585	2.03
	30	105	1780	2.65

Source: Adapted from Huang *et al.* (2010). Reproduced with permission of Springer.

heuristic function is informed and this usually happens at the bottom of the search graph (Pearl, 1984). Thus, Algorithm 10.1 that employs A* locally in all stages of the search performs better than the BF-BT algorithm that only employs A* at the top of the search graph. Second, the important decisions with respect to the quality of a schedule may happen at early stages of the search (Xiong & Zhou, 1998). Thus, Algorithm 10.1 can decrease the likelihood of missing critical candidates when compared with the BT-BF algorithm that employs a backtracking search at early stages.

2) If $N_{max} = 1$, Algorithm 10.1 does not degenerate to a pure backtracking algorithm and it outperforms the backtracking method in terms of the makespan of the obtained schedule. The reason is that these two algorithms use different strategies to select a state for expansion at each iteration. In Algorithm 10.1 with $N_{max} = 1$, if a state is chosen for expansion at an iteration, all of its enabled transitions fire to generate its successors. Then, the best state with the minimal

Table 10.3 Scheduling results of the SC-net in Figure 9.3 with the lot size (10, 10, 6, 6).

Algorithm	$d_{max}(N_{max})$	c	N_E	$T(s)$
A*	—	134	23 092	720.00
BT	—	274	193	0.38
	80	206	6 281	64.00
	100	198	12 341	240.00
BF-BT	120	180	16 602	480.00
	140	169	20 155	540.00
	160	153	21 797	660.00
	80	209	1 254	5.00
	100	181	2 315	16.00
BT-BF	120	162	8 495	139.00
	140	150	11 368	390.00
	160	148	18 875	560.00
	1	170	581	0.62
	5	155	809	0.93
	10	146	1 089	1.25
Algorithm 10.1	15	147	1 538	2.50
	20	144	1 889	2.81
	25	142	2 850	5.62
	30	141	2 995	7.03

Source: Adapted from Huang *et al.* (2010). Reproduced with permission of Springer.

f-value among these successors is chosen for the next expansion. However, in a backtracking method, if a state is chosen for expansion, only one enabled transition fires to generate an immediate succeeding state. If the successor is not a goal state, then it is expanded at the next iteration. Therefore, the quality of a state chosen for expansion at each iteration in Algorithm 10.1 with $N_{max} = 1$ may be better than that of the backtracking method in terms of f-values.

10.3 Illustrative Examples

In this section, we test Algorithm 10.1 with more complex problems having the following characteristics: i) buffers with limited sizes, ii) alternative operations, and iii) operations with multiple kinds of resources. These characteristics are shown in Figure 10.6 where $o_{i,j,k}$ represents the jth operation of the ith job requiring the

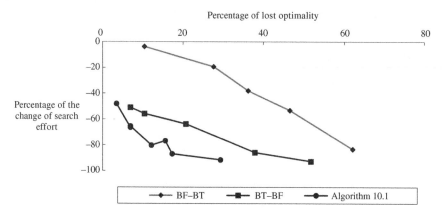

Figure 10.3 $\mathcal{R}_E$ versus $\mathcal{R}_c$ for the SC-net with the lot size $(5, 5, 2, 2)$. *Source:* Adapted from Huang *et al.* (2010). Reproduced with permission of Springer.

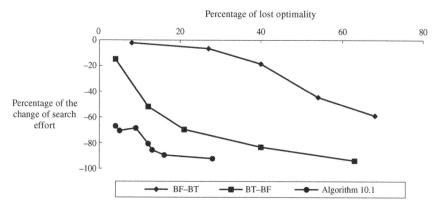

Figure 10.4 $\mathcal{R}_E$ versus $\mathcal{R}_c$ for the SC-net with the lot size $(8, 8, 4, 4)$. *Source:* Adapted from Huang *et al.* (2010). Reproduced with permission of Springer.

kth resource. The size of a buffer can be modeled as Figure 10.6a. It is represented by the number of tokens in the place denoting the buffer size. For example, the buffer size of the net shown in Figure 10.6a is three. Figure 10.6b represents that the jth operation of the ith job can be performed through an alternative operation, i.e., by using either operation $o_{i,j,k}$ with resource k or operation $o_{i,j,r}$ with r. Figure 10.6c shows that the modeled operation requires two kinds of resources, i.e., R_k and R_r. We randomly generate 40 RAS problems that are generated by randomly selecting and linking predefined Petri net modules (Mejía, 2002). They have the following characteristics. Each problem has three kinds of shared resources and four different jobs. Each job requires three operations to finish its tasks. 75% of the operations have an alternative operation and 40% of the operations require

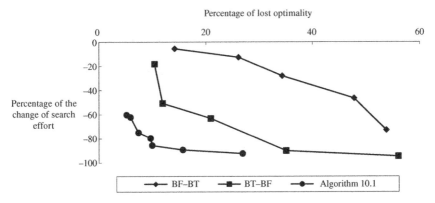

Figure 10.5 $\mathcal{R}_E$ versus $\mathcal{R}_c$ for the SC-net with the lot size $(10, 10, 6, 6)$. Source: Adapted from Huang *et al.* (2010). Reproduced with permission of Springer.

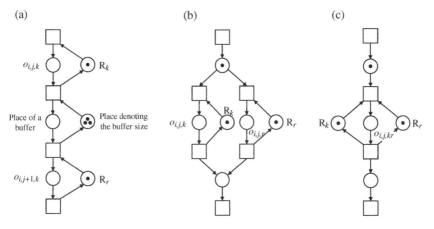

Figure 10.6 (a) A buffer with a finite size; (b) an operation with an alternative operation; and (c) an operation that requires two resources. Source: Adapted from Huang *et al.* (2010). Reproduced with permission of Springer.

two kinds of resources. Each operation has been assigned a random operation time from a uniform distribution [1, 100]. The size of each buffer is randomly selected from 1 to 3 and the lot size of each job is also randomly selected from 1 to 3.

We solve the set of random problems by using Algorithm 10.1 with the value of $N_{\max}$ in $\{1, 5, 10, 15, ..., 60, +\infty\}$. Algorithm 10.1 with $N_{\max} = +\infty$ equals a pure A* algorithm. The comparisons of the scheduling results are summarized in Figure 10.7. The upper curve denotes the mean percentages of the lost optimality $\mathcal{R}_c$ and the lower curve represents the mean percentages of the change in search effort $\mathcal{R}_E$ for different values of $N_{\max}$. We see that Algorithm 10.1 can

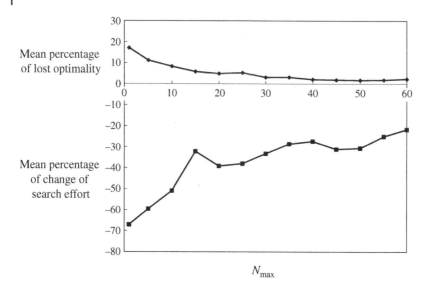

Figure 10.7 Performances of Algorithm 10.1 for random problems. Source: Adapted from Huang *et al.* (2010). Reproduced with permission of Springer.

greatly reduce the computational complexity in terms of the number of expanded states and lower the mean lost optimality when compared with the A* search. For instance, Algorithm 10.1 with $N_{max} = 30$ explores 33% fewer states than the A* search (i.e., Algorithm 10.1 with $N_{max} = +\infty$), but its $\mathcal{R}_c$ is only about 3% on average.

10.4 Concluding Remarks

SC-nets can well model RASs with concurrency, synchronization, sequencing, resource sharing, and operation times. So, they are an ideal and popular tool to model and schedule these systems. The reachability graph of an SC-net represents the whole state space of its underlying system. Lee and DiCesare (1994) pioneered the combination of the reachability graph and a heuristic A* algorithm to schedule such systems. It only needs to explore a partial reachability graph to obtain an optimal schedule with an admissible heuristic function. However, a problem observed in the method is the state explosion problem, which means that, although the exploration of the whole reachability graph is avoided, the number of the expanded states still grows exponentially with the problem size and/or initial marking. It makes the method only applicable to small systems. To speed up the pathfinding in the reachability graph, this chapter presents a hybrid search strategy that performs an A* search locally and a backtracking search globally in the reachability graph of an SC-net to efficiently search for a near-optimal schedule for RASs.

10.5 Bibliographical Notes

This chapter presents a hybrid heuristic algorithm to quickly schedule RASs based on place-timed Petri nets. The material of the presented method mainly comes from Huang *et al.* (2010). The classifications and applications of different timed Petri nets can be seen in Wang (1998). Recent results about time Petri nets can be found from (Pan *et al.*, 2020 and applications from (Bai *et al.*, 2016; Qiao *et al.*, 2019; Yang *et al.*, 2020). For more details about the modeling of discrete event systems based on Petri nets, please refer to Hrúz and Zhou (2007), Seatzu *et al.* (2013), Wu and Zhou (2010), and Zhou and DiCesare (1993). Other intelligent search algorithms can be found in Edelkamp and Schroedl (2012), Pearl (1984), and Russell and Norvig (2010).

11

A* Search with More Informed Heuristics Functions

To efficiently schedule resource allocation systems (RASs), this chapter presents and evaluates another improved search method in an SC-net framework. Within the reachability graphs of place-timed SC-nets for RASs, the method can simultaneously use admissible and inadmissible heuristic functions in A* search to find optimal schedules for the underlying systems. It also proves that the resultant combinational heuristic function is still admissible and more informed than any of its constituents. An example system and several sets of randomly generated problems are tested to show the effectiveness and efficiency of the presented method.

11.1 Introduction

Petri nets have often been used as a tool to model and schedule RASs (Baruwa et al., 2015; Qiao et al., 2018; Tuncel & Bayhan, 2007; Yang et al., 2017; Zhu et al., 2016, 2018). To find optimal or suboptimal schedules for RASs, Lee and DiCesare (1994) combine Petri net simulation capabilities with the A* search algorithm, which uses a heuristic function to only expand the most promising branches of the reachability graphs. If an admissible heuristic function is used, the method can guarantee the optimality of the obtained result. In Lee and Lee (2010), Mejía (2002), Moro et al. (2002b), Xiong and Zhou (1998), and Yu et al. (2003), some admissible heuristic functions for the A* search in the reachability graphs have been proposed. However, a typical problem of the method is that the A* algorithm with an admissible heuristic function often costs much time to find an optimal solution even for a medium-size problem. To reduce such search effort, Lee and Lee (2010) and Mejía (2002) develop some inadmissible heuristic functions that invoke quicker termination conditions at the cost of the results' optimality.

This chapter presents a method for the A* algorithm within reachability graphs of SC-nets to simultaneously use several heuristic functions no matter whether they are admissible or not for RAS scheduling. The procedures of how to combine

Supervisory Control and Scheduling of Resource Allocation Systems: Reachability Graph Perspective, First Edition. Bo Huang and MengChu Zhou.

admissible heuristics and inadmissible heuristics are presented and the properties of the combinational heuristic are given. It ensures that the combinational heuristic is not only admissible but also more informed than its constituents. The method has the same application scope as the method in Chapter 8, that is to say, it can be applied to the classes of Petri nets for RASs including PPN, S^3PR, ES^3PR, LS^3PR, ELS^3PR, GLS^3PR, WS^3PR, S^3PGR^2, WS^3PSR, S^4R, S^4PR, G-system, S^*PR, S^5PR, and PC^2R (the details of these nets can be seen in Section 2.2.5) with minor revisions shown in Section 8.2.2 and the SC-nets constructed by the methods in Section 8.2.3.

11.2 More Informed Heuristics in A* Search

To schedule RASs, the A* algorithm is often used to find good executable schedules. Its detailed procedures can be seen in Algorithm 8.1. In the A* algorithm, its used heuristic function h is critical for both search speed and result's quality. In this section, we give some definitions and properties of heuristic functions used in the A* algorithm. Let $(\mathcal{N}, S_0)$ be a net system of an SC-net of RASs and $G(\mathcal{N}, S_0)$ be its reachability graph.

Definition 11.1 *In an A* algorithm, if a heuristic function h is always a lower bound to all solutions descending from the current state S to the goal state S_G, i.e., $h(S) \leq h^*(S), \forall S \in G(\mathcal{N}, S_0)$ where $h^*(S)$ denotes the optimal cost going from S to S_G, then the heuristic function h is said to be admissible, which guarantees that the result obtained by the A* algorithm is optimal, i.e., the obtained schedule has the minimal cost.*

Definition 11.2 *Given a heuristic function h, if there exists a state S in $G(\mathcal{N}, S_0)$ such that $h(S) > h^*(S)$, h is said to be inadmissible.*

Note that an A* algorithm with an inadmissible heuristic function does not ensure the optimality of results.

Definition 11.3 *Given two admissible heuristic functions h_1 and h_2, h_2 is said to be more informed than h_1 if $h_2(S) > h_1(S)$ for every nongoal state S. An A* algorithm using h_2 is said to be more informed than that using h_1.*

Definition 11.4 *A heuristic function h is said to be monotone or consistent if $h(S) \leq c(S, S') + h(S'), \forall(S, S') \in G(\mathcal{N}, S_0)$ and $h(S_G) = 0$.*

Monotonicity is a condition that is slightly stronger than admissibility, i.e., a monotone heuristic function is also admissible. The following theorems about monotone heuristic functions come from Edelkamp & Schroedl (2012).

Theorem 11.1 *The A* algorithm guided by a monotone heuristic function finds optimal paths to all expanded states, i.e., $g(S) = g^*(S)$, $\forall S \in CLOSED$ where $g^*(S)$ denotes the lowest cost among all paths from the initial state S_0 to S.*

Theorem 11.2 *Monotonicity implies that the f-values of a sequence of states expanded by A* is non-decreasing.*

11.3 Combination of Admissible and Inadmissible Heuristics

In this section, the definitions of relative error, upper bound, and informedness of a heuristic function of A* are presented. Then, a method is given to use both admissible and inadmissible heuristic functions that have upper bounds of negative relative errors for the A* search in reachability graphs. Finally, the proofs of that the combinational heuristic function is admissible and more informed than any of its constituents are given.

A relative error is the magnitude of the difference between an exact value and an approximation divided by the magnitude of the exact value. Analogously, the definition of the relative error of a heuristic function can be given below.

Definition 11.5 *Given a state S, the relative error of h at S is defined as*

$$e = \left| \frac{h^*(S) - h(S)}{h^*(S)} \right|. \tag{11.1}$$

If $h^*(S) > h(S)$, we say that there is a positive relative error of h at S, which is denoted as

$$e^+ = \frac{h^*(S) - h(S)}{h^*(S)}; \tag{11.2}$$

if $h^*(S) < h(S)$, we say that there is a negative relative error of h at S, which is denoted as

$$e^- = \frac{h(S) - h^*(S)}{h^*(S)}. \tag{11.3}$$

Property 11.1 *If h is an admissible heuristic function, then a positive relative error e^+ exists for h at any state $S \in G(\mathcal{N}, S_0)$; if h is an inadmissible heuristic function, then a negative relative error e^- of h exists for at least one state $S \in G(\mathcal{N}, S_0)$.*

Let $\Pr(X \leq x)$ be the probability that X is not greater than X. The definition of an upper bound of a variable can be given as follows.

Definition 11.6 *A random variable X is said to have an upper bound s if $X \leq s$ and $\forall x < s$, $\Pr(X \leq x) < 1$.*

By using Definition 11.6, an upper bound of the negative relative errors of an inadmissible heuristic h can be defined below.

Definition 11.7 *Let h be an inadmissible heuristic function. The negative relative error e^- of h at any reachable state is said to have an upper bound s if $\forall S$, $\frac{h(S)-h^*(S)}{h^*(S)} \leq s$ and $\forall x < s$, $Pr\left(\frac{h(S)-h^*(S)}{h^*(S)} \leq x\right) < 1$.*

For an A* algorithm, its used heuristic function is very important since an admissible heuristic function guarantees the optimality of the obtained results and has a pruning power in the search process, i.e., it makes the A* algorithm only expand the most promising branches of the reachability graphs. Some admissible heuristic functions have been proposed in the literature for RAS scheduling based on reachability graphs (Lee & Lee, 2010; Mejía, 2002; Moro *et al.*, 2002*b*; Xiong & Zhou, 1998; Yu *et al.*, 2003). For admissible heuristic functions, the informedness in Definition 11.3, which indicates the accuracy of the estimates and the pruning power of an admissible heuristic function, can be used to compare their quality. However, Definition 11.3 is too strict to be satisfied in practice. By following (Pearl, 1984), a relaxed version of more informedness of an admissible heuristic function is given as follows.

Definition 11.8 *If the same tie-breaking rule that is independent of g-values and h-values is used, then the definition of more informedness of an admissible heuristic function h_1 when compared with another admissible function h_2 can be relaxed to permit equalities between them, i.e., $h_1(S) \geq h_2(S)$ for every nongoal state S.*

Although the A* search with an admissible heuristic function can ensure the optimality of its obtained results, the admissibility of a heuristic function is often too strict for many problems, that is to say, one cannot always design an admissible heuristic function for an A* application. In addition, the A* search with an admissible heuristic still suffers from the state explosion problem, i.e., the number of explored states increases exponentially with the problem size. Thus, some inadmissible heuristics are also proposed and used in the literature (Lee & DiCesare, 1994; Lee & Lee, 2010; Mejía, 2002) to expedite the A* search process at the cost of the result's quality. Based on these different kinds of heuristic functions, this section presents a method that combines admissible heuristic functions with inadmissible ones for the A* algorithm and proves that the combinational heuristic function is admissible and more informed than all of its constituents.

Theorem 11.3 *Given a set of heuristic functions $h_{A_1}, \ldots, h_{A_l}, h_{\bar{A}_{l+1}}, \ldots, h_{\bar{A}_k}$ in which $h_{A_1}, \ldots, h_{A_l}$ are admissible and $h_{\bar{A}_{l+1}}, \ldots, h_{\bar{A}_k}$ are inadmissible. Let $s_{\bar{A}_{l+1}}, \ldots, s_{\bar{A}_k}$ be* *the upper bounds of the negative relative errors of* $h_{\bar{A}_{i+1}}, \dots, h_{\bar{A}_k}$, *respectively. The combinational heuristic function*

$$h_{COM} = \max \left\{ h_{A_1}, \dots, h_{A_i}, \frac{h_{\bar{A}_{i+1}}}{1 + s_{\bar{A}_{i+1}}}, \dots, \frac{h_{\bar{A}_k}}{1 + s_{\bar{A}_k}} \right\} \tag{11.4}$$

is admissible and more informed than any of its constituents.

Proof: *We first prove that* $\frac{h_{\bar{A}_{i+1}}}{1+s_{\bar{A}_{i+1}}}, \dots, \frac{h_{\bar{A}_k}}{1+s_{\bar{A}_k}}$ *are admissible heuristic functions. Since* $s_{\bar{A}_{i+1}}, \dots, s_{\bar{A}_k}$ *are the upper bounds of the negative relative errors of the inadmissible functions* $h_{\bar{A}_{i+1}}, \dots, h_{\bar{A}_k}$, *respectively, we have that for any reachable state S,* $e_j^- = \frac{h_{\bar{A}_j}(S) - h^*(S)}{h^*(S)} \leqslant s_{\bar{A}_j}$ ($j = i + 1, \dots, k$) *according to Definition 11.7. So,* $\forall S$, $\frac{h_{\bar{A}_j}(S)}{1+s_{\bar{A}_j}} \leqslant \frac{h_{\bar{A}_j}(S)}{1+e_j^-} = h^*(S)$. *Thus,* $\frac{h_{\bar{A}_j}}{1+s_{\bar{A}_j}}$ ($j = i+1, \dots, k$) *is admissible.*

In addition, $h_{A_1}, \dots, h_{A_i}$ *are admissible. It implies that for any reachable state S,* $h_{A_1}(S) \leqslant h^*(S), \dots, h_{A_i}(S) \leqslant h^*(S)$. *Thus, we have* $\forall S$,

$$h_{COM}(S) = \max \left\{ h_{A_1}(S), \dots, h_{A_i}(S), \frac{h_{\bar{A}_{i+1}}(S)}{1 + s_{\bar{A}_{i+1}}}, \dots, \frac{h_{\bar{A}_k}(S)}{1 + s_{\bar{A}_k}} \right\} \leq h^*(S).$$

Therefore, h_{COM} *is admissible. According to Definition 11.8,* h_{COM} *is more informed than any of its constituents since* h_{COM} *and its constituents are all admissible and the max operation renders* h_{COM} *greater than or equal to any of its constituents at any reachable state.* ∎

11.4 Illustrative Examples

This section gives the performance tests of the combinational heuristic function developed in this chapter. First, we give an upper bound of the negative relative errors of the inadmissible heuristic h'_{LBR} proposed in Lee and Lee (2010). Next, we combine the inadmissible heuristic h'_{LBR} with some admissible heuristic functions such as h_ξ (Xiong & Zhou, 1998) and h_{RCR} (Moro *et al.*, 2002b; Yu *et al.*, 2003) by using Theorem 11.3 to generate an admissible and more informed heuristic function. Note that h_ξ is a special case of h_H which is presented in Section 9.3 because h_ξ does not consider alternative routes, weighted arcs, multiple copies of resources, or the tokens remaining times in the system models. Then an example from Xiong and Zhou (1998) with different lot sizes is tested by using the A* searches with different heuristic functions. Finally, some random scheduling problems are also generated and tested to demonstrate the method presented in this chapter. In the search process of A*, the used tie-breaking rule is last-in-first-out, which is independent of the g-values and h-values of the states.

In Lee and Lee (2010), an inadmissible heuristic h'_{LBR} for the A* search is proposed as follows:

$$h'_{LBR}(S) = \max_i \text{Dist}(S_i, S_{G_i}) \tag{11.5}$$

where $\text{Dist}(S_i, S_{G_i})$ denotes the minimal processing time required to move all parts in the ith job from S_i to S_{G_i}. S_i (respectively, S_{G_i}) denotes the current state (respectively, goal state) in the ith job. In the following, we give an upper bound of the negative relative errors of h'_{LBR}.

Theorem 11.4 *An upper bound of the negative relative errors of h'_{LBR} is $s_{LBR} = l_{maxDist} - 1$ where $l_{maxDist}$ is the number of parts in the job that has the maximal value of $\text{Dist}(S_i, S_{G_i})$ at state S.*

Proof: According to the definitions given in (Lee & Lee, 2010), $h'_{LBR}(S)$ is the maximal processing time required by any job to move all parts of the job from S_i to S_{G_i} along a path that needs the smallest processing time. Suppose that all parts in any job at S can be simultaneously processed and finished. Then, the processing time obtained from dividing h'_{LBR} by $l_{maxDist}$ is less than or equal to h^* since the supposed situation is an extreme case in which the needed processing time from S to S_G is minimal. So, we have $\forall S, \frac{h'_{LBR}(S)}{l_{maxDist}} \leq h^*(S)$. Hence, $\forall S, \frac{h'_{LBR}(S) - h^*(S)}{h^*(S)} = \frac{h'_{LBR}(S)}{h^*(S)} - 1 \leq l_{maxDist} - 1$.

Then, we suppose that, at any state S, all parts in the job that has the maximal value of $\text{Dist}(S_i, S_{G_i})$ can be simultaneously processed and finished. We have $\frac{h'_{LBR}(S)}{l_{maxDist}} = h^*(S)$, which means that the processing time needed by the state transfer from S to S_G is minimal. So, in this case, the negative relative error of h'_{LBR} at S is

$$e^- = \frac{h'_{LBR}(S) - h^*(S)}{h^*(S)} = l_{maxDist} - 1. \tag{11.6}$$

However, in any other cases, the system may spend more processing time on the state transfer from S to S_G, i.e., $\frac{h'_{LBR}(S)}{l_{maxDist}} > h^*(S)$. So, we have $\forall S, \frac{h'_{LBR}(S) - h^*(S)}{h^*(S)} \leq l_{maxDist} - 1$ and $\forall x < l_{maxDist} - 1$

$$\text{Pr}\left(\frac{h'_{LBR}(S) - h^*(S)}{h^*(S)} \leq x\right) < 1. \tag{11.7}$$

By Definition 11.7, $s_{LBR} = l_{maxDist} - 1$ is an upper bound of the negative relative error of h'_{LBR}. ∎

In Eq. (11.8), a new heuristic function h_{COM} is formulated based on the admissible heuristics h_{ξ} and h_{RCR} and the inadmissible heuristic h'_{LBR} according to Theorem 11.3. It is guaranteed that h_{COM} is admissible and more informed than any of its constituents by the theorem.

$$h_{\mathrm{COM}}(S) = \max \left\{ h_{\xi}(S), h_{\mathrm{RCR}}(S), \frac{h'_{\mathrm{LBR}}(S)}{1 + s_{\mathrm{LBR}}} \right\}$$

$$= \max \left\{ h_{\xi}(S), h_{\mathrm{RCR}}(S), \frac{h'_{\mathrm{LBR}}(S)}{l_{\mathrm{maxDist}}} \right\}. \tag{11.8}$$

To demonstrate the presented method, we compare the performances of the A* algorithms with different heuristics in terms of the makespan c of the obtained solution, number of expanded states N_E, and computational time $\mathcal{T}$. First, we consider the Petri net model of a manufacturing system in Xiong and Zhou (1998) with different lot sizes $(1, 1, 1, 1), (2, 2, 1, 1), (5, 5, 2, 2), (8, 8, 4, 4)$, and $(10, 10, 6, 6)$. The scheduling results are shown in Tables 11.1–11.5. We can see that the A* algorithm with the presented heuristic function h_{COM} obtains the same makespans as the A* searches with the admissible heuristics (h_{RCR} and h_{ξ}) and it is more efficient than the A* search with any of its heuristic constituents ($h'_{\mathrm{LBR}}/l_{\mathrm{maxDist}}$, h_{RCR} and h_{ξ}) in

Table 11.1 Performance comparisons for the system of Xiong and Zhou (1998) with lot size $(1, 1, 1, 1)$.

Heuristic	c	N_E	$\mathcal{T}$ (s)
$h'_{\mathrm{LBR}}(S)$	17	598	139
$h'_{\mathrm{LBR}}(S)/l_{\mathrm{maxDist}}$	17	598	156
$h_{\mathrm{RCR}}(S)$	17	1573	407
$h_{\xi}(S)$	17	41	2
$h_{\mathrm{COM}}(S)$	17	34	2

Source: Adapted from Huang *et al.* (2014). Reproduced with permission of Elsevier.

Table 11.2 Performance comparisons for the system of Xiong and Zhou (1998) with lot size $(2, 2, 1, 1)$.

Heuristic	c	N_E	$\mathcal{T}$ (s)
$h'_{\mathrm{LBR}}(S)$	25	37 892	165 170
$h'_{\mathrm{LBR}}(S)/l_{\mathrm{maxDist}}$	25	58 127	520 104
$h_{\mathrm{RCR}}(S)$	25	54 612	468 330
$h_{\xi}(S)$	25	394	31
$h_{\mathrm{COM}}(S)$	25	189	14

Source: Adapted from Huang *et al.* (2014). Reproduced with permission of Elsevier.

Table 11.3 Performance comparisons for the system of Xiong and Zhou (1998) with lot size $(5, 5, 2, 2)$.

Heuristic	c	N_E	$\mathcal{T}$ (s)
$h'_{\text{LBR}}(S)$	—	—	—
$h'_{\text{LBR}}(S)/l_{\text{maxDist}}$	—	—	—
$h_{\text{RCR}}(S)$	—	—	—
$h_\varsigma(S)$	58	974	122
$h_{\text{COM}}(S)$	58	868	113

Source: Adapted from Huang *et al.* (2014). Reproduced with permission of Elsevier.

Table 11.4 Performance comparisons for the system of Xiong and Zhou (1998) with lot size $(8, 8, 4, 4)$.

Heuristic	c	N_E	$\mathcal{T}$ (s)
$h'_{\text{LBR}}(S)$	—	—	—
$h'_{\text{LBR}}(S)/l_{\text{maxDist}}$	—	—	—
$h_{\text{RCR}}(S)$	—	—	—
$h_\varsigma(S)$	100	1838	443
$h_{\text{COM}}(S)$	100	1685	403

Source: Adapted from Huang *et al.* (2014). Reproduced with permission of Elsevier.

Table 11.5 Performance comparisons for the system of Xiong and Zhou (1998) with lot size $(10, 10, 6, 6)$.

Heuristic	c	N_E	$\mathcal{T}$ (s)
$h'_{\text{LBR}}(S)$	—	—	—
$h'_{\text{LBR}}(S)/l_{\text{maxDist}}$	—	—	—
$h_{\text{RCR}}(S)$	—	—	—
$h_\varsigma(S)$	134	2567	736
$h_{\text{COM}}(S)$	134	2507	731

Source: Adapted from Huang *et al.* (2014). Reproduced with permission of Elsevier.

terms of N_E and $\mathcal{T}$ for the cases with lot sizes $(1, 1, 1, 1)$ and $(2, 2, 1, 1)$. For the cases with lot sizes $(5, 5, 2, 2)$, $(8, 8, 4, 4)$, and $(10, 10, 6, 6)$, both the A* algorithm with h_{COM} and the A* algorithm with h_ξ can terminate with a result in a reasonable amount of time. In addition, the A* search with h_{COM} is more efficient than the A* search with h_ξ in terms of N_E and $\mathcal{T}$.

Next, we generate some SC-nets for RASs by randomly selecting and linking predefined Petri net modules (Mejía & Odrey, 2005) and putting operation times on activity places. The number of resources m, the number of jobs n, the lot size l_i of a part type P_i, the number of plans q_i of P_i, the number of tasks u_{ij} of a plan P_{ij}, the number of alternative operations v_{ijk} of a task T_{ijk}, the number of resources s_{ijkl} required by an operation o_{ijkl}, and the operation time d_{ijkl} of an operation o_{ijkl} are all randomized. All random values are generated from a uniform distribution between one and a predefined maximal integer. The predefined maximal integers of m, n, l_i, q_i, u_{ij}, v_{ijk}, s_{ijkl}, and d_{ijkl} are denoted by $\hat{M}$, $\hat{N}$, $\hat{L}$, $\hat{Q}$, $\hat{U}$, $\hat{V}$, $\hat{S}$, and $\hat{D}$, respectively.

First, we randomly generate 50 small-scale models. Their scheduling results are shown in Table 11.6 where the relative difference between the makespans obtained by two Algorithms a and b is defined as

$$\mathcal{R}_c(a, b) = \frac{c(a) - c(b)}{c(b)} \times 100\%, \tag{11.9}$$

where $c(x)$ denotes the makespan obtained by an algorithm x. In the sequel, we use a to denote the A* algorithm with a tested heuristic function and b to denote the A* search with $h = 0$ that can obtain a schedule with the minimal makespan.

Table 11.6 Performance comparisons for small-scale problems ($\hat{M} = \hat{S} = \hat{L} = \hat{N} = \hat{U} = \hat{Q} = \hat{V} = 2$, $\hat{D} = 50$, number of problems = 50).

Heuristic	Average $\mathcal{R}_c(a, b)$	Average N_E	Average $\mathcal{T}$ (s)
$h'_{LBR}(S)$	0.15	14.34	0.44
$h'_{LBR}(S)/l_{maxDist}$	0.00	18.02	0.56
$h_{RCR}(S)$	0.00	10.98	0.40
$h_\xi(S)$	0.00	7.04	0.30
$h_{COM}(S)$	0.00	6.92	0.58

Different cases in the calculation of h_{COM}	Average times
$h_{COM}(S) = h'_{LBR}(S)/l_{maxDist}$	4.86
$h_{COM}(S) = h_{RCR}(S)$	6.72
$h_{COM}(S) = h_\xi(S)$	10.82

Source: Adapted from Huang *et al.* (2014). Reproduced with permission of Elsevier.

We see that, for the 50 small-scale problems, the average relative difference R_c for the A* search with h_{COM} is zero. The reason is that h_{COM} is an admissible heuristic function and it guarantees the optimality of the results obtained by an A* algorithm with the heuristic function. For these small-scale problems, in terms of the number of expanded states N_E, the A* search with h_{COM} expands fewer states than the A* searches with its heuristic constituents. However, the computational time $\mathcal{T}$ of the A* with h_{COM} is a little bigger than the A* searches with its heuristic constituents. This is probably because the A* with h_{COM} needs extra time to compute all of its constituent heuristic functions for each generated state in the calculation of h_{COM}. In Table 11.6, "average times" denotes the average number of occurrences of each case. We can see that, for the A* algorithm with h_{COM}, each of its heuristic constituents has the opportunity to be the largest constituent that is selected as h_{COM} at some states. Thus, h_{COM} can take full advantage of different heuristic functions and becomes more informed than its constituents.

Tables 11.7 and 11.8 show the experimental results for two sets of 50 middle-scale problems. The results in Table 11.7 are similar to those for the small-scale problems in Table 11.6 except that the A* search with h_{COM} becomes more efficient than its constituents in terms of the computational time. This phenomenon can be explained by considering that, as the problem scale increases, the saved time by expanding fewer states with a more informed heuristic h_{COM} increases and exceeds the extra time required to compute all constituent heuristics of h_{COM} for each generated state. From the results in Table 11.8, we can see that when both $\hat{M}$ and $\hat{S}$ increase to 6 and $\hat{N}$ and $\hat{U}$ increase to 4 and 3, respectively, both

Table 11.7 Performance comparisons for middle-scale problems I ($\hat{M} = \hat{S} = 4$, $\hat{N} = \hat{U} = 3$, $\hat{L} = \hat{Q} = \hat{V} = 2$, $\hat{D} = 50$, number of problems = 50).

Heuristic	Average $R_c(a,b)$	Average N_E	Average $\mathcal{T}$ (s)
$h'_{LBR}(S)$	1.35	358.18	129.34
$h'_{LBR}(S)/l_{maxDist}$	0.00	377.40	128.44
$h_{RCR}(S)$	0.00	328.04	133.48
$h_g(S)$	0.00	55.72	5.24
$h_{COM}(S)$	0.00	50.56	4.52

Different cases in the calculation of h_{COM}	Average times
$h_{COM}(S) = h'_{LBR}(S)/l_{maxDist}$	13.50
$h_{COM}(S) = h_{RCR}(S)$	9.30
$h_{COM}(S) = h_g(S)$	100.12

Source: Adapted from Huang *et al.* (2014). Reproduced with permission of Elsevier.

Table 11.8 Performance comparisons for middle-scale problems II ($\hat{M} = \hat{S} = 6$, $\hat{N} = 4$, $\hat{U} = 3$, $\hat{L} = \hat{Q} = \hat{V} = 2$, $\hat{D} = 50$, number of problems = 50).

Heuristic	Average $R_c(a, b)$	Average N_E	Average T (s)
$h'_{\text{LBR}}(S)$	0.04	4357.04	39 507.48
$h'_{\text{LBR}}(S)/l_{\text{maxDist}}$	0.00	3030.36	38 024.14
$h_{\text{RCR}}(S)$	0.00	2560.38	32 272.92
$h_\varsigma(S)$	0.00	213.20	98.46
$h_{\text{COM}}(S)$	0.00	192.80	81.86

Different cases in the calculation of h_{COM}	Average times
$h_{\text{COM}}(S) = h'_{\text{LBR}}(S)/l_{\text{maxDist}}$	51.18
$h_{\text{COM}}(S) = h_{\text{RCR}}(S)$	13.30
$h_{\text{COM}}(S) = h_\varsigma(S)$	469.26

Source: Adapted from Huang *et al.* (2014). Reproduced with permission of Elsevier.

the number of expanded states N_E and the computational time T of the A* algorithms with h'_{LBR}, $\dfrac{h'_{\text{LBR}}}{l_{\text{maxDist}}}$, and h_{RCR} increase dramatically. However, the average relative difference R_c between the makespans obtained by A* with $h = h_{\text{COM}}$ and A* with $h = 0$ is still zero. In addition, the A* algorithm with the presented heuristic h_{COM} is more efficient than the A* algorithms with $\dfrac{h'_{\text{LBR}}}{l_{\text{maxDist}}}$, h_{RCR}, and h_ς in terms of average N_E and T.

From the results in these tables, we conclude that it is promising to use the presented heuristic function h_{COM} as a candidate heuristic function for the A* search within the reachability graphs of SC-nets for RAS scheduling. The combinational heuristic function h_{COM} can take full advantage of its constituent heuristics, no matter whether they are admissible or not, and it is still admissible and more informed than its constituents, making the A* search run faster.

11.5 Concluding Remarks

To speed up the RAS scheduling, this chapter presents a strategy to design a new heuristic function by simultaneously using existing admissible and inadmissible heuristic functions for the A* search within the reachability graphs of SC-nets for RASs if an upper bound of negative relative errors of each inadmissible heuristic function can be found. It also proves that the combinational heuristic function is still admissible and more informed than its constituents. Thus, the A* search with the combinational heuristic function can obtain an optimal schedule faster than the A* searches with those existing heuristics. Finally, five cases of a manufacturing system with different lot sizes are tested by using the A* search with different heuristics. Many randomly generated problems with different scale are also tested to show the effectiveness and efficiency of the presented method.

In the future, this method can be extended in the following aspects. First, it is challenging to formulate more informed heuristic functions to schedule large-scale RAS problems. Second, setting different performance indices such as minimization of tardiness and developing multicriteria heuristic searches are interesting issues.

11.6   Bibliographical Notes

The heuristic function used in an A* algorithm is very important to RAS scheduling in terms of its search speed and solution' quality. A historical progression of the applications of Petri nets to the scheduling of systems can be found in Tuncel and Bayhan (2007). A survey and classification of A*-based heuristic searches can be seen in Rios & Chaimowicz (2010). For the properties of a heuristic A* search, please refer to the books (Edelkamp & Schroedl, 2012; Pearl, 1984; Russell & Norvig, 2010). The method reported in this chapter on combining different kinds of heuristic functions to design an admissible and more informed heuristic function for an A* search in reachability graphs mainly comes from Huang *et al.* (2014).

12

Symbolic Heuristic Search

To fast schedule resource allocation systems (RASs), this chapter presents a symbolic A* search approach to compactly represent and efficiently manipulate SC-nets of RASs. First, RASs are modeled by SC-nets where time information is associated with activity places. Then, the procedures to functionally represent, evolve, and schedule these timed SC-nets are given by combining binary decision diagrams and the A* search. An admissible heuristic function suitable for the symbolic method and its efficient functional implementation are also proposed. When compared with the existing explicit-state A* methods, the symbolic method can compactly represent large sets of states of timed Petri nets and efficiently manipulate such sets to fast compute optimal schedules for the underlying systems. Several experiments are conducted to show the effectiveness and efficiency of the presented method.

12.1 Introduction

Since Petri nets (PNs) can well model and analyze RASs with concurrency, synchronization, sequencing, and resource sharing (Hrúz & Zhou, 2007), they are an ideal and popular tool to handle RAS scheduling problems. Lee and DiCesare (1994) pioneered in combining Petri nets' reachability graphs with the A* search to schedule RASs. By using an admissible heuristic function, their method only needs to explore a partial reachability graph to find an optimal schedule from an initial marking to a goal marking if such a path exists. But it suffers from the state explosion problem (Pearl, 1984; Russell & Norvig, 2010), which means that although the generation of the whole reachability graph is avoided by the A* search, the number of the explored markings grows exponentially with the problem size. It makes the algorithm only applicable to small systems. To speed up its search process in reachability graphs, a lot of effort has been made to alleviate the problem, for example, A* algorithms with controlled or limited backtracking strategies (Lefebvre, 2016;

Supervisory Control and Scheduling of Resource Allocation Systems: Reachability Graph Perspective,
First Edition. Bo Huang and MengChu Zhou.
© 2020 The Institute of Electrical and Electronics Engineers, Inc. Published 2020 by John Wiley & Sons, Inc.

Lei *et al.*, 2017; Mejía, 2002; Peng *et al.*, 2019; Xiong & Zhou, 1998), A* searches with limited local search scope (Moro *et al.*, 2002b; Luo *et al.*, 2015; Mejía & Niño, 2017), real-time A* searches that use a rule-based supervisor to reduce the search space (Kim *et al.*, 2007), and anytime heuristic searches that combine iterative deepening A* with breadth-first heuristic search and backtracking (Baruwa *et al.*, 2015). We have also developed some methods to accelerate the search process, such as hybrid heuristic A* (Huang *et al.*, 2008, 2010) and dynamic weighted A* (Huang *et al.*, 2012a). However, these methods accelerate the A* search at the cost of the results' optimality, i.e., the optimality of the obtained schedules cannot be guaranteed and sometimes the schedule's quality may be poor.

A Binary Decision Diagram (BDD) is a data structure used to represent a Boolean function (Xu *et al.*, 2016). It is a directed acyclic graph that can represent a large set of encoded data with a small data structure and allow efficient computations on such sets. For untimed PNs, Pastor *et al.* (2001) use BDDs to quickly generate reachable markings and analyze PN's structural and behavioral properties (such as liveness and concurrency). BDDs are also used to efficiently compute legal markings (Chen *et al.*, 2011) and minimal siphons (Chen & Liu, 2013) of untimed PNs. However, time information is necesasry in RAS scheduling and the method that accelerates the A* search by combining BDDs and timed PNs has not been investigated.

This chapter presents a symbolic method that combines the BDD representations with timed SC-nets to efficiently schedule timed PNs of RASs. First, RASs are modeled by SC-nets in which time information is associated with activity places. Then, the state representation and model evolution of timed SC-nets are symbolically formulated by using BDD characteristic functions. Based on these Boolean representations, a symbolic A* search and its admissible heuristic function are designed to obtain optimal schedules for RASs. The symbolic method uses the compressed data structures of BDDs to reason about sets of states instead of individual markings and executes an efficient functional exploration in the SC-nets' reachability graphs. The results of some experiments show that the method is much more efficient than the explicit-state A* search and it is a promising way to alleviate the state explosion problem in RAS scheduling.

12.2 Boolean Algebra and Binary Decision Diagram

This section briefly reviews the basic concepts of Boolean algebra and BDDs. Boolean algebra is defined as $(B, +, \cdot, 0, 1)$ in which B is a set called the carrier whose elements satisfy the commutative, distributive, identity, and complement laws, + (Logic OR) and · (Logic AND) are binary operations on B, and 0 and 1 are elements of B. Let E be a set and 2^E be the power set of E. The algebra

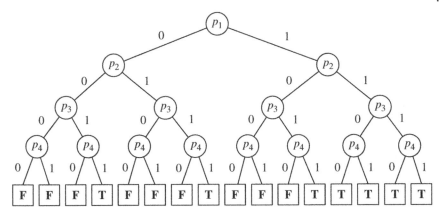

Figure 12.1 An ordered BDD representing a set of markings.

of 2^E, denoted as $(2^E, \bigcup, \bigcap, \emptyset, E)$ where $\bigcup$, $\bigcap$, and $\emptyset$ are the union operation, intersection operation, and the empty set, respectively, is Boolean algebra.

Let l be a non-negative integer and $\mathcal{F}$ be an l-variable Boolean function from B^l to B. Let $\mathcal{F}_l(B)$ be the set of Boolean functions from B^l to B. The algebra of $\mathcal{F}_l(B)$, denoted as $(\mathcal{F}_l(B), +, \cdot, \mathbf{0}, \mathbf{1})$ where $\mathbf{0}$ and $\mathbf{1}$ denote the functions $\mathcal{F}(x_1, \ldots, x_l) = 0$ and $\mathcal{F}(x_1, \ldots, x_l) = 1$, respectively, is Boolean algebra.

A BDD is a rooted, directed, acyclic graph with two terminal nodes labeled "F" (called bddfalse) and "T" (called bddtrue). In a BDD, the non-terminal nodes are labeled with Boolean variables x_i, $i \in \{1, \ldots, n\}$, and the edges among the nodes are labeled with either 1 or 0. The term BDD usually refers to reduced and ordered BDD in which no two non-terminal nodes exist with the same 1-edge and 0-edge and a fixed ordering of variables is preserved in all paths of BDD. Such BDD is unique for a particular Boolean function and variable order.

Figure 12.1 shows an ordered BDD that represents a set of markings $\{p_3 + p_4, p_2 + p_3 + p_4, p_1 + p_3 + p_4, p_1 + p_2, p_1 + p_2 + p_3, p_1 + p_2 + p_4, p_1 + p_2 + p_3 + p_4\}$ that have the formula $(p_1 \wedge p_2) \vee (p_3 \wedge p_4)$. Figure 12.2 gives its reduced form that represents the same set of markings. A marking is in the set if and only if the marking's path from the root node down through the BDD ends with a "T" node. In the sequel, "BDD" is used to denote "reduced and ordered BDD."

12.3 Symbolic Evolution of Place-Timed Petri Nets

A resource allocation system can be modeled by a place-timed SC-net via a conversion method from an existing untimed Petri net as in Section 8.2.2 or a top-down or bottom-up modeling method as in Section 8.2.3. This section shows how to symbolically represent and evolve a place-timed SC-net via BDDs.

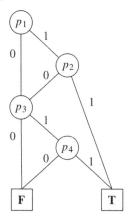

Figure 12.2 A reduced and ordered BDD representing the same set of markings.

First, the Boolean representations of relations between two non-negative integer variables are introduced. Let a and b be two non-negative integer variables that are, respectively, represented by two sets of Boolean variables $a^0, a^1, \ldots, a^{k_a}$ and $b^0, b^1, \ldots, b^{k_b}$ with $k_a = k_b$. The relation $a = b$ can be given as:

$$(a^{k_a} \equiv b^{k_a}) \cdot (a^{k_a - 1} \equiv b^{k_a - 1}) \cdots (a^1 \equiv b^1) \cdot (a^0 \equiv b^0)$$

$$= \prod_{i=0}^{k_a} (a^i \equiv b^i). \tag{12.1}$$

Similarly, $a \geq b$ can be represented as:

$$(a^{k_a} > b^{k_a})$$
$$+ (a^{k_a} \equiv b^{k_a}) \cdot (a^{k_a - 1} > b^{k_a - 1})$$
$$\ldots\ldots$$
$$+ (a^{k_a} \equiv b^{k_a}) \cdot (a^{k_a - 1} \equiv b^{k_a - 1}) \cdots (a^1 \equiv b^1) \cdots (a^0 > b^0)$$
$$+ (a^{k_a} \equiv b^{k_a}) \cdot (a^{k_a - 1} \equiv b^{k_a - 1}) \cdots (a^1 \equiv b^1) \cdot (a^0 \equiv b^0)$$

$$= \sum_{i=0}^{k_a} \left[(a^i > b^i) \cdot \prod_{j=i+1}^{k_a} (a^j \equiv b^j) \right] + \prod_{i=0}^{k_a} (a^i \equiv b^i). \tag{12.2}$$

For an untimed PN $N = (P, T, F, W)$, the number of tokens $M(p)$ in a place $p \in P$ at a marking M, which is up to k, can be represented at least $k_p + 1$ of Boolean variables $p^0, p^1, \ldots,$ and p^{k_p}. The fewest Boolean variables required for such a k-bounded place is $k_p + 1 = \lceil \log_2(k+1) \rceil$. For example, if p_1 is 2-bounded, $M(p_1)$ can be represented by two Boolean variables p_1^1 and p_1^0. $p_1^1 p_1^0$ denotes $M(p_1) = 3$, $p_1^1 \overline{p}_1^0$ denotes $M(p_1) = 2$, $\overline{p}_1^1 p_1^0$ denotes $M(p_1) = 1$, and $\overline{p}_1^1 \overline{p}_1^0$ denotes $M(p_1) = 0$ where $\overline{p}$ denotes applying a NOT operator to a Boolean variable. Similarly, for an arc $(x, y) \in F$ in N, its weight $W(x, y)$ can be represented by the same number of Boolean variables $w^0, w^1, \ldots,$ and w^{k_p} since an arc weight is often not greater than the maximal token capacity of a place that connects the arc. Based on these notations, the condition under which a transition t is enabled at a marking M can be given by the following characteristic function, denoted as $\mathcal{E}_t$:

$$\prod_{p \in \bullet t} [M(p) \geq W(p, t)]$$

$$= \prod_{p \in \bullet t} \left[\sum_{i=0}^{k_p} \left[(p^i > w^i) \cdot \prod_{j=i+1}^{k_p} (p^j \equiv w^j) \right] + \prod_{i=0}^{k_p} (p^i \equiv w^i) \right]. \quad (12.3)$$

However, (12.3) cannot ensure the enabling of a transition for a timed SC-net $\mathcal{N}$. In an SC-net $\mathcal{N}$, operation times are associated with activity places. Thus, a token in an activity place p may indicate: i) this token is already available for the enabling and firing of a post-transition of p; or ii) although the token is not yet ready for enabling and firing a transition, it should become available after a fixed time. Note that, in the second case, remaining time exists for the token to be available. As time passes by, the token's remaining time decreases until it reaches zero. When the remaining time is zero, the token is available for its corresponding place's output transitions' enabling and firing. Thus, the tokens' remaining times are as important as the tokens' distribution for a timed PN $\mathcal{N}$ to successfully "reach" a system state. Therefore, both marking and remaining time matrix are required to denote a state in $\mathcal{N}$.

Suppose that activity places in $\mathcal{N}$ are κ-bounded. Let $R_u: P_A \to \mathbb{N}, u \in \{1, \ldots, \kappa\}$ be a vector of the remaining times of the uth token in each activity place. Then, $S = (M, R_1, \ldots, R_\kappa)$ can be used to represent a state of $\mathcal{N}$. For example, an initial state of $\mathcal{N}$ can be denoted as $S_0 = (M_0, [0], \ldots, [0])$ in which no tokens exist in any activity place at the initial state.

Since SC-nets include time information, the enabling condition of a transition in an SC-net $\mathcal{N}$ is different from that in an untimed PN. Let $G(\mathcal{N}, S_0)$ be the reachability graph of an SC-net $\mathcal{N}$ with S_0. In $G(\mathcal{N}, S_0)$, an edge (S, S') leads the system from S to S' by the firing of a transition t. However, in an activity pre-place p_i of t at the marking of S, a token required by the enabling of t may not be available since the token must stay in p_i for at least the time $D(p_i)$. So, the cost of an edge (S, S') in $G(\mathcal{N}, S_0)$, denoted as $c(S, S')$ that refers to the time needed by the state transfer from S to S', can be intuitively considered as the maximal operation time among the activity pre-places of t, i.e.,

$$c(S, S') = \max_{p_i \in \bullet t \cap P_A} \{D(p_i)\}. \quad (12.4)$$

But this intuitive argument is not quite correct. In the state transfer from S to S', tokens in all places are concurrently delayed by the time obtained by (12.4). It is true for each state update. Furthermore, in a timed SC-net $\mathcal{N}$, we have $(P_A \times T) \cup (T \times P_A) \rightarrow \{0, 1\}$, i.e., the firing of t needs exactly one token from each activity pre-place of t. Such a token has the least remaining time among the tokens in the same place. So, the time spent on the state transfer from S to S' is not the maximal operation time of activity pre-places of t, but rather the maximum of the least remaining time of tokens in activity pre-places of t, i.e.,

$$c(S, S') = \max_{p_i \in {}^\bullet t \cap P_A} \left\{ \min_{u \in \{1, \dots, \kappa\}} \{R_{i,u}\} \right\} \tag{12.5}$$

where $R_{i,u}$ denotes $R_u(p_i)$ for simplicity. Then, the condition in which a transition t satisfying (12.3) is not enabled at a state of $\mathcal{N}$ due to the token availability can be given by the following characteristic function, denoted as $\tilde{F}_t$:

$$\sum_{p_i \in {}^\bullet t \cap P_A} \sum_{v=1}^{\kappa} \left[[M(p_i) = v] \cdot \prod_{u=1}^{v} [R_{i,u} \geq 1] \right]. \tag{12.6}$$

This condition means that, for a transition t in $\mathcal{N}$, there exists at least one activity pre-place of t that cannot provide an available token for the enabling of t. Thus, if both (12.3) and (12.6) are satisfied, we must wait until the remaining times of the required tokens decrease to zero to enable a transition t in $\mathcal{N}$. Algorithm 12.1 gives the implementation of such a process in the form of Boolean manipulations.

Algorithm 12.1

```
Decreasing remaining times of tokens in activity places to enable a
transition at a state of an SC-net.
Input: A transition t and a BDD S representing a set of states
satisfying (12.3).
Output: A BDD representing a set of states in which t is enabled.
Function Dₜ(S) {
        C₁ := S · F̃ₜ;  /*By (12.6)*/
        Z := S \ C₁;
        while C₁ ≠ bddfalse
        {
                for each i satisfying pᵢ ∈ Pₐ and D(pᵢ) > 0
                {
                        for u := 1 to κ
                        {
                                C₂  :=  C₁ · (R_{i,u} ≥ 1);
                                if C₂ ≠ bddfalse
                                {
                                        C₁ := C₁ \ C₂;
                                                  k_{r_{i,u}}
                                        C₂ := ∃_{r_{i,u}}  ∏  [s_{i,u}ʲ ≡ ηʲ(r_{i,u})] · C₂;
                                                  j:=0
                                        C₂[s ↔ r];
                                        C₁ := C₁ + C₂;
                                }
                        }
```

$$
\begin{aligned}
&\quad\quad \} \\
&\quad \} \\
&\quad \mathcal{Z} \;:=\; \mathcal{Z} + \left[C_1 \setminus (C_1 \cdot \tilde{F}_t) \right]; \\
&\quad C_1 \;:=\; C_1 \cdot \tilde{F}_t; \\
&\} \\
&\texttt{return } \mathcal{Z};
\end{aligned}
$$

}

In Algorithm 12.1, $[s \leftrightarrow r]$ denotes the swapping of two sets of variables to allow multiple quantifications, $R_{i,u}$ is denoted by a set of Boolean variables $r^0_{i,u}$, $r^1_{i,u}$, ..., and $r^{k_{r_{i,u}}}_{i,u}$, and η represents the following symbolic function to subtract one from $R_{i,u}$:

$$
\eta^j(r_{i,u}) = \begin{cases} \bar{r}^j_{i,u}, & \text{if } j = 0 \\ r^j_{i,u} \oplus \mathbb{B}_{j-1}, & \text{if } j \geq 1 \end{cases} \tag{12.7}
$$

where $\oplus$ represents the bitwise operation of exclusive OR and $\mathbb{B}_j$ denotes a borrow function given as follows:

$$
\mathbb{B}_j = \begin{cases} \bar{r}^j_{i,u}, & \text{if } j = 0 \\ \bar{r}^j_{i,u} \cdot \mathbb{B}_{j-1}, & \text{if } j \geq 1. \end{cases} \tag{12.8}
$$

Algorithm 12.1 obtains a BDD representing a set of states at which t is enabled for both tokens' distribution and token availability. If there exists a state where a token required to enable t is unavailable, the algorithm decreases the remaining times of all tokens in activity places by one at each iteration. The process continues until all the tokens required to enable t are available.

Next, we consider the Boolean representations of the token movement if an enabled transition is selected to fire in $\mathcal{N}$. When a transition t fires, for a place $p_i \in P$, the transformation of the $(j+1)$th Boolean variable of $M(p_i)$ can be given as follows:

$$
\delta^{t^i}(p_i) = \begin{cases} p^j_i \oplus \omega^j \oplus \mathbb{B}'_{j-1}, & \text{if } p_i \in {}^\bullet t \\ p^j_i \oplus \omega^j \oplus \mathbb{C}_{j-1}, & \text{if } p_i \in t^\bullet \\ p^j_i, & \text{otherwise} \end{cases} \tag{12.9}
$$

where $\mathbb{B}'_j$, $\mathbb{C}'_j$, and ω^j are, respectively, the following borrow, carry, and arc weight transform functions:

$$
\mathbb{B}'_j = \begin{cases} \bar{p}^j_i \cdot \omega^j + \mathbb{B}'_{j-1} \cdot (p^j_i \equiv \omega^j), & \text{if } j \geq 0 \\ 0, & \text{otherwise} \end{cases} \tag{12.10}
$$

$$
\mathbb{C}'_j = \begin{cases} p^j_i \cdot \omega^j + \mathbb{C}'_{j-1} \cdot (p^j_i \oplus \omega^j), & \text{if } j \geq 0 \\ 0, & \text{otherwise} \end{cases} \tag{12.11}
$$

$$
\omega^j = \begin{cases} W^j(p_i, t), & \text{if } p_i \in {}^\bullet t \\ W^j(t, p_i), & \text{if } p_i \in t^\bullet. \end{cases} \tag{12.12}
$$

Then, the relation of token distributions before and after the firing of t can be given as:

$$R_t(p_1, \ldots, p_m, q_1, \ldots, q_m) = \prod_{i=1}^{m} \prod_{j=0}^{k_{p_i}} \left[q_i^j \equiv \delta^{t^j}(p_i) \right] \tag{12.13}$$

where $m = |P|$. After tokens move, newly added tokens in activity post-places of t are associated with their corresponding operation times. The Boolean manipulations of this process are given in Algorithm 12.2.

Algorithm 12.2

```
Adding operation times to new tokens in activity post-places of t.
Input: A transition t and a BDD S representing a set of states.
Output: A BDD representing a set of states with operation times on
the new tokens in t,s activity post-places.
Function A_t(S) {
        for each i satisfying p_i ∈ t• ∩ P_A  and  D(p_i) > 0
        {
            for u := κ to 1
            {
                C_1 := [R_{i,u} = 0];
                C_2 := bddtrue;
                for v := 1 to u − 1
                    C_2 := C_2 · [R_{i,v} ⩾ 1];
                Z := S · C_1 · C_2;
                if Z ≠ bddfalse
                {
                    S := S \ Z;
                    for j := 0 to k_{r_{i,u}}
                        Z|_{r'_{i,u} := D(p_i)^j}   ; /*Substitution by the process-
ing time of p_i.*/
                        S := S + Z;
                }
            }
        }
        return S;
}
```

In Algorithm 12.2, if an post-place of t, which is an activity place, has received a new token due to the firing of t, the token is given the operation time of the activity place as its remaining processing time. In Algorithm 12.2, the operation time is associated with the first zero entry in $R_{i,u}$. Note that the presented algorithm performs it on sets of states, not on individual states or markings, by using BDD manipulations.

Finally, a mapping function that expands a set of states to generate successors by firing a transition t in $\mathcal{N}$ can be given below:

$\text{Img}(\mathcal{N}, S)$

$$= \exists_{p_1,\dots,p_m} \sum_{t \in T} A_t \left(D_t(S \cdot \mathcal{E}_t) \cdot \prod_{i=1}^{m} \prod_{j=0}^{k_{p_i}} \left[q_i^j \equiv \delta^{t^j}(p_i) \right] [q \leftrightarrow p] \right) \qquad (12.14)$$

where S is a BDD representing a set of states and $[q \leftrightarrow p]$ denotes the swap of two sets of variables to allow multiple quantifications after the token movement.

12.4 Symbolic Heuristic Search

This section presents the steps of a symbolic A* search method for the scheduling of place-timed SC-nets. An admissible heuristic function for the symbolic search and its Boolean implementation are also given.

In the symbolic A* search for a place-timed SC-net $\mathcal{N} = (P, T, F, W, D)$, the aforementioned state is extended as $S = (f, M, R_1, \dots, R_\kappa)$ with $f(S) = g(S) + h(S)$ where $f(S)$ denotes an estimate of the cost or makespan from the initial state S_0 to a given goal state S_G along an optimal path that goes through the current state S, $g(S)$ represents the current lowest cost from S_0 to S, and $h(S)$ is a heuristic estimate of the cost from S to S_G along an optimal path. Let $S.X$ be an element X of S. A transition $t \in T$ is enabled at a state S if $\forall p \in {}^\bullet t$, $S \cdot M(p) \geq W(p, t)$ and $\forall p \in {}^\bullet t \cap P_A$, p has at least one token whose remaining time is zero. Firing an enabled transition t generates a new state S' such that i) $S' \cdot M[t\rangle S \cdot M$; ii) $g(S') = g(S) + c(S, S')$; iii) $\forall p_i \in P$, $\forall u \in \{1, \dots, \lambda\}$, $S' \cdot R_{i,u} = \max\{0, S \cdot R_{i,u} - c(S, S')\}$; and iv) $\forall p \in t^\bullet \cap P_A$, a new token exists in p at S' such that its remaining time is $D(p)$. This firing process is denoted by $S[t\rangle S'$ where S is called an immediate predecessor of S' and S' is called an immediate successor of S in its reachability graph $G(\mathcal{N}, S_0)$. A state S' is called reachable from S_0 if there exists a sequence of transition firings from S_0 to S' in $G(\mathcal{N}, S_0)$. The symbolic A* algorithm for $\mathcal{N}$ is given presented in Algorithm 12.3, which is different from the explicit-state A* in Lee and DiCesare (1994) since the symbolic A* adopts BDDs to compactly represent sets of states and efficiently operate on them for the A* heuristic search within the reachability graphs of place-timed PNs.

Algorithm 12.3

```
Symbolic A* algorithm for place-timed PNs.
Input: An SC-net N, an initial state S₀, and a goal state S_G.
Output: A transition firing sequence from S₀ to S_G if it exists.
Astar_search {
```

$$Open(S) := (f = S_0.f) \cdot \prod_{p_i \in P} [M(p_i) = S_0.M(p_i)] \cdot \prod_{p_i \in P_A} \prod_{u:=1}^{\kappa} (R_{i,u} = S_0.R_{i,u});$$

```
Closed(S) := bddfalse;
loop
{
    if Open(S) = bddfalse
        return ''Exploration completed with failure!'';
    f_min := min{f'|Open(S)·(f = f') ≠ bddfalse};
    Front(S) := Open(S)·(f = f_min);
    if Front(S)· ∏ [M(p_i) = S_G.M(p_i)] ≠ bddfalse
              p_i∈P
        return Construct(Closed(S),S_0,S_G);
    Succ(S) := Img(PN, Front(S));
    Closed(S) := Closed(S) + Front(S);
    Succ(S) := Succ(S) \ Closed(S);
    Open(S) := Open(S) \ Front(S);
    Open(S) := Open(S) + Succ(S);
}
}
```

In Algorithm 12.3, the *Open* BDD represents the set of states already generated but not yet expanded, while the *Closed* BDD represents the set of states that have been expanded. A BDD evaluates to bddtrue for a given state if and only if that state belongs to the set. At each iteration of Algorithm 12.3, immediate successors of the states that have the minimal f-value in *Open* are generated by using the Boolean mapping operation in (12.14). If S_G is expanded, the algorithm terminates and a path from S_0 to S_G can be extracted as follows. A state that comes before S_G in the path must be contained in *Closed*. Therefore, by intersecting the predecessors of S_G with *Closed*, each state in the intersection is in the path, so all such states can be obtained to continue solution reconstruction until S_0 is found. Hence, by iteratively keeping track of such predecessors in the *Closed* BDD, a sequence of states from *Closed* and their transition firings (actions) linking S_0 to S_G in $G(\mathcal{N}, S_0)$ can be obtained.

The optimality of the symbolic A* is inherited from the fact that an explicit-state A* with an admissible heuristic function guarantees the optimality of the obtained schedule. However, the heuristic functions used in Huang *et al.* (2014), Lee and DiCesare (1994), Lee and Lee (2010), Luo *et al.* (2015), Mejía (2002), Moro *et al.* (2002b), Peng *et al.* (2019), and Xiong and Zhou (1998) are not suitable for Algorithm 12.3 since these functions are designed for individual markings. In the following, we present a heuristic function suitable for the symbolic A* in Algorithm 12.3. It can operate on BDDs that represent sets of states. In addition, its admissibility is proven and its symbolic implementation is given.

Given a resource place $p_i \in P_R$, the activity places whose operations require p_i in any path from S_0 to S_G are called the loyal holders of p_i, denoted as $H_l(p_i)$. In addition, the total time to be spent on $H_l(p_i)$ from S_0 to S_G is called the loyal time of p_i, denoted as $\mathcal{T}_{H_l(p_i)}$, i.e.,

$$\mathcal{T}_{H_l(p_i)} = \sum_{p \in H_l(p_i)} [D(p) \cdot M_0(p_s)] \tag{12.15}$$

where p_s is the start place of the processing subnet containing p. Let P_R^1 be the set of 1-bounded resource places and p_{max} be a resource place whose loyal time is maximal among the places in P_R^1. Then, a heuristic function for the symbolic A* algorithm can be given as follows:

$$h_{\mathrm{BDD}}(S) = \begin{cases} 0, & \text{if } P_R^1 = \emptyset \\ \mathcal{T}_{H_l(p_{max})}, & \text{if } P_R^1 \neq \emptyset \text{ and } S = S_0 \\ h_{\mathrm{BDD}}(S') - \tau_{p_{max}}(S', S), & \text{otherwise} \end{cases} \quad (12.16)$$

where S' is an immediate predecessor of S and $\tau_{p_{max}}(S', S)$ is the time spent on $H_l(p_{max})$ in the state transfer from S' to S. That is to say, $h_{\mathrm{BDD}}(S)$ is the total processing time of operations that are definitely to be performed with p_{max} from S to S_G.

We use the net on the right of Figure 8.2 as an example to illurstrate it. Suppose that S_0 is the initial state where all part tokens are in start places and S_G is the goal state where all part tokens are in end places. The loyal holders of each resource place are $H_l(p_{14}) = \emptyset$, $H_l(p_{15}) = \{p_{12}\}$, $H_l(p_{16}) = \{p_6\}$, $H_l(p_{17}) = \{p_{10}\}$, $H_l(p_{18}) = \{p_2, p_5, p_{11}, p_{13}\}$, and $H_l(p_{19}) = \{p_7, p_9\}$. Note that p_4 is not a loyal holder of p_{14} since there exists an alternative operation (denoted by p_3), which means that the operation of p_4 with resource p_{14} is not necessary for the state transfer from S_0 to S_G. Then, we have $p_{max} = p_{18}$ since the loyal time of p_{18}, i.e., $\mathcal{T}_{H_l(p_{18})} = D(p_2) \cdot M_0(p_1) + D(p_5) \cdot M_0(p_1) + D(p_{11}) \cdot M_0(p_8) + D(p_{13}) \cdot M_0(p_8) = 96$, is maximal among the places in $P_R^1 = \{p_{14} - p_{18}\}$. Thus, the heuristic function for this net is

$$h_{\mathrm{BDD}}(S) = \begin{cases} 96, & \text{if } S = S_0 \\ h_{\mathrm{BDD}}(S') - \tau_{p_{max}}(S', S), & \text{otherwise.} \end{cases} \quad (12.17)$$

Next, the admissibility of h_{BDD} is proven.

Theorem 12.1 h_{BDD} *is admissible.*

Proof: *If $P_R^1 = \emptyset$, h_{BDD} is admissible by Definition 2.35. So, we should only consider the case when $P_R^1 \neq \emptyset$. First, we prove that $\forall S \in R(\mathcal{N}, S_0), \forall S'' \in R(\mathcal{N}, S)$,*

$$h_{BDD}(S) \leq c^*(S, S'') + h_{BDD}(S'') \quad (12.18)$$

where $R(\mathcal{N}, S_0)$ is the set of reachable states in $G(\mathcal{N}, S_0)$ and $c^*(S, S'')$ is the cost or time spent on the best path from S to S'' in $G(\mathcal{N}, S_0)$. It is proven by induction. $\forall S \in R(\mathcal{N}, S_0), \forall S'' \in R(\mathcal{N}, S)$, if S'' is an immediate successor of S, then $h_{BDD}(S) = \tau_{p_{max}}(S, S'') + h_{BDD}(S'') \leq c^*(S, S'') + h_{BDD}(S'')$ since $\tau_{p_{max}}(S, S'')$ is the time to be spent on the loyal holders of p_{max} in the state transfer from S to S''. Suppose that if S'' is a depth-n successor of S, $h_{BDD}(S) \leq c^*(S, S'') + h_{BDD}(S'')$ holds. If S'' is a depth-$(n+1)$ successor of S, we have

$$h_{BDD}(S) = \tau_{p_{max}}(S, S''') + h_{BDD}(S''')$$
$$\leq c^*(S, S''') + h_{BDD}(S''')$$
$$\leq c^*(S, S''') + c^*(S''', S'') + h_{BDD}(S'')$$
$$\leq c^*(S, S'') + h_{BDD}(S'')$$

where S''' is an immediate successor of S in a path from S to S'' in $G(\mathcal{N}, S_0)$. So, (12.18) is confirmed. Then, we replace S'' in (12.18) with S_G and obtain $\forall S \in R(\mathcal{N}, S_0)$, $h_{BDD}(S) \leq c^*(S, S_G) + h_{BDD}(S_G)$. Now, since $h_{BDD}(S_G) = 0$ and $c^*(S, S_G) = h^*_{BDD}(S)$, we have $h_{BDD}(S) \leq h^*_{BDD}(S)$. Therefore, h_{BDD} is admissible. ∎

Therefore, according to the optimality of an A* algorithm, if Algorithm 12.3 uses h_{BDD}, it can terminate with an optimal solution when a solution exists. In the following, we show that if h_{BDD} is used, the computation of the g-values and h-values are avoided in the calculations of the f-values of newly generated states. For any state S', we have $f(S') = g(S') + h_{BDD}(S')$. Since we can access the f-value of S', but usually not its g-value and h-value, the f-value of an immediate successor S of S' can be computed as follows:

$$f(S) = g(S) + h_{BDD}(S) = g(S') + c(S', S) + h_{BDD}(S)$$
$$= f(S') + c(S', S) + h_{BDD}(S) - h_{BDD}(S')$$
$$= f(S') + c(S', S) - \tau_{p_{max}}(S', S). \qquad (12.19)$$

It means that if h_{BDD} is used, we do not need to compute the g-values and h-values of new states to obtain their f-values. Instead, we only need to update their f-values via (12.19). More importantly, (12.19) can be implemented by slightly revising D_t in Algorithm 12.1. Algorithm 12.4 presents the revised D_t that decreases tokens' remaining times to enable a transition and updates f-values of the newly generated states.

Algorithm 12.4

```
Decreasing the remaining times of tokens in activity places to
enable a transition and updating f-values.
Input: A transition t and a BDD S representing a set of states
satisfying (12.3).
Output: A BDD representing a set of states at which t is enabled
and f-values are updated for t,s firing.
Function Dₜ(S) {
     C₁ := S · F̃ₜ;
     Z := S \ C₁;
     while C₁ ≠ bddfalse
     {
          for each i satisfying pᵢ ∈ Pₐ and D(pᵢ) > 0
          {
```

```
for u := 1 to κ
{
        C₂ := C₁ · (R_{i,u} ≥ 1);
        if C₂ ≠ bddfalse
        {
                C₁ := C₁ \ C₂;
                                    k_{r_{i,u}}
                C₂ := ∃_{r_{i,u}} ∏    [s^j_{i,u} ≡ η^j(r_{i,u})] · C₂;
                                    j:=0
                C₂[s ↔ r];
                if p_i ∈ H_l(p_max)
                {
                        /*Subtract 1 from f.*/
                                        k_f
                        C₂ := ∃_f ∏   [e^j ≡ η^j(f)] · C₂;
                                        j:=0
                        C₂[e ↔ f];
                }
                C₁ :=  C₁ + C₂;
        }
}
                k_f
C₁ := ∃_f ∏   [e^j ≡ α^j(f)] · C₁;  /*Add 1 to f.*/
                j:=0
C₁[e ↔ f];
Z := Z + [C₁ \ (C₁ · F̃_l)];
C₁ := C₁ · F̃_l;
}
return Z;
}
```

In Algorithm 12.4, the f-value of a state is represented by a set of Boolean variables $f^0, f^1, \ldots,$ and f^{k_f} and α is a symbolic function to add 1 to f, which is defined as:

$$\alpha^j(f) = \begin{cases} \overline{f^j}, & \text{if } j = 0 \\ f^j \oplus C_{j-1}, & \text{if } j \geq 1 \end{cases} \tag{12.20}$$

where C_j is the following carry function:

$$C_j = \begin{cases} f^j, & \text{if } j = 0 \\ f^j \cdot C_{j-1}, & \text{if } j \geq 1. \end{cases} \tag{12.21}$$

Note that since p_{max} is 1-bounded, at most one marked loyal holder of p_{max} exists at any reachable state. So, $\tau_{p_{max}}(S', S)$ equals the time spent on a marked loyal holder (if it exists) of p_{max} in the state transfer from S' to S. Thus, according to (12.19), whenever $\tau_{p_{max}}(S', S)$ increases by one in Algorithm 12.4, the f-value of S should decrease by one. In addition, as the remaining times of tokens gradually decrease by one at each iteration, the f-value of S increases by one. Note that these operations are performed on BDDs that represent sets of states in the presented symbolic method.

12.5 Illustrative Examples

In this section, some example systems are tested with the symbolic A* search within the reachability graphs of timed SC-nets. For the application of BDDs, C programs have been developed with the BDD tool Buddy-2.2 package (Lind-Nielsen, 2002) on a Linux computer with an 8 Gb memory and an Intel i5 Core 2.2 Ghz CPU. The variables of states are symbolically encoded in an interleaved way, that is, if two variables are needed and they are represented by the Boolean variables, a^0, a^1 and b^0, b^1, respectively, then the order of Boolean variables is a^0, b^0, a^1, and b^1.

First, a flexible manufacturing system in Uzam (2002) and Li *et al.* (2008a) is considered. It has six resources R_1–R_6 that may be machines, robots, etc., two input buffers I_1 and I_2, and two output buffers O_1 and O_2. Two types of parts, P_1 and P_2, are processed with the following processing routings:

$$P_1: I_1 \rightarrow R_3 \rightarrow R_2(\text{or } R_5) \rightarrow R_3 \rightarrow R_6 \rightarrow R_4 \rightarrow O_1$$
$$P_2: I_2 \rightarrow R_4 \rightarrow R_1 \rightarrow R_3 \rightarrow R_2 \rightarrow R_3 \rightarrow O_2$$

Let $o_{i,j,k}$ be the *j*th operation of the *i*th job, which requires the *k*th resource. The operation times of the system are given in Table 12.1. Its SC-net model is shown in Figure 12.3 that has 14 transitions and 21 places in which $P_S = \{p_1, p_8\}$, $P_E = \{p_{20}, p_{21}\}$, $P_R = \{p_{14} - p_{19}\}$, and the rest belong to the activity places P_A. Eight cases with different lot sizes are tested by the explicit-state A* and the presented symbolic A*. The makespan of the obtained schedule, the number of expanded states (N_E), the number of BDD nodes for the expanded states (N_{BDD}), and the computational time ($\mathcal{T}$) are shown in Table 12.2. From the results, the following conclusions are delivered:

1) The schedules obtained by the symbolic A* with $h = h_{BDD}$ are guaranteed to be optimal since h_{BDD} is admissible.

Table 12.1 Operation times of Example 1.

Operation	Operation time	Operation	Operation time
$o_{1,1,3}$	3	$o_{2,1,4}$	2
$o_{1,2,2}$	2	$o_{2,2,1}$	4
$o_{1,2,5}$	4	$o_{2,3,3}$	4
$o_{1,3,3}$	4	$o_{2,4,2}$	3
$o_{1,4,6}$	3	$o_{2,5,3}$	5
$o_{1,5,4}$	5		

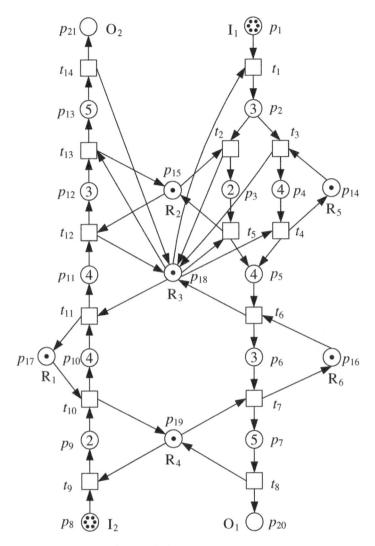

Figure 12.3 SC-net of Example 1.

2) For the same system, the symbolic A* expands more states than the explicit-state A*. Such a phenomenon is expected due to the following two facts: First, the symbolic A* efficiently performs on sets of states, while the explicit-state A* executes on individual states; Second, when new states are generated, the symbolic A* allows to insert the states that are not in *Closed* into *Open* for expansion for computational efficiency, while the explicit-state A* spends time checking each state and picking promising states into *Open*.

Table 12.2 Scheduling results for Example 1.

p_1	p_8	Makespan	Explicit A* ($h = 0$)		Symbolic A* ($h = 0$)			Symbolic A* ($h = h_{BDD}$)		
			N_E	$T(s)$	N_E	N_{BDD}	$T(s)$	N_E	N_{BDD}	$T(s)$
1	1	21	5.60×10^1	0.01	2.04×10^2	9.23×10^3	0.55	9.60×10^1	5.64×10^3	0.71
2	2	35	1.05×10^2	0.16	1.41×10^4	1.63×10^5	4.62	1.10×10^3	2.77×10^4	1.08
3	3	51	6.17×10^3	4.65	1.45×10^5	6.73×10^5	40.50	6.18×10^3	1.05×10^5	3.11
4	4	67	1.85×10^4	50.18	5.68×10^5	1.37×10^6	137.67	2.18×10^4	2.50×10^5	9.37
5	5	83	3.47×10^4	194.99	1.43×10^6	2.23×10^6	292.97	4.95×10^4	4.05×10^5	17.96
6	6	99	5.32×10^4	463.03	2.86×10^6	3.13×10^6	460.59	8.90×10^4	5.32×10^5	27.29
7	7	115	7.50×10^4	943.46	4.99×10^6	3.76×10^6	686.53	1.40×10^5	6.32×10^5	35.49
8	8	131	1.01×10^5	1718.39	7.95×10^6	4.23×10^6	911.54	2.03×10^5	6.99×10^5	46.05

Table 12.3 Operation times of Example 2.

Operation	Operation time	Operation	Operation time
$O_{1,1,1}$	5	$O_{3,1,3}$	5
$O_{1,2,2}$	4	$O_{3,2,1}$	2
$O_{1,3,3}$	4	$O_{3,3,2}$	5
$O_{2,1,2}$	2	$O_{4,1,3}$	2
$O_{2,2,3}$	5	$O_{4,2,2}$	4
$O_{2,3,1}$	2	$O_{4,3,1}$	2

3) By using BDD's compact Boolean representations and efficient logical operations, the symbolic A* is much faster than the explicit-state A* to schedule the place-timed SC-nets, especially for larger cases and with an informed heuristic function $h = h_{BDD}$.

The second example system has an intermediate buffer between any two consecutive operations to hold parts that are ready for the next operation. The system consists of four input buffers I_1–I_4, four output buffers O_1–O_4, and three resources R_1–R_3. Four types of parts, P_1–P_4, are processed in the system. The processing routes are given below:

$$P_1: I_1 \rightarrow R_1 \rightarrow R_2 \rightarrow R_3 \rightarrow O_1$$
$$P_2: I_2 \rightarrow R_2 \rightarrow R_3 \rightarrow R_1 \rightarrow O_2$$
$$P_3: I_3 \rightarrow R_3 \rightarrow R_1 \rightarrow R_2 \rightarrow O_3$$
$$P_4: I_4 \rightarrow R_3 \rightarrow R_2 \rightarrow R_1 \rightarrow O_4$$

The operation times of the system are shown in Table 12.3. The SC-net of the system is given in Figure 12.4 that has 24 transitions and 31 places in which $P_S = \{p_1, p_8, p_{15}, p_{22}\}$, $P_E = \{p_7, p_{14}, p_{21}, p_{28}\}$, and $P_R = \{p_{29}, p_{30}, p_{31}\}$, while the rest belong to the activity places P_A. Note that, the activity places without time information represent the intermediate buffers and the other activity places with a non-negative integer (i.e., an operation time) represent operations to be performed in the system with specific resources. For this net, five cases with different lot sizes are tested by using explicit-state and symbolic A*. The scheduling results are given in Table 12.4. It can be seen that these methods obtain results with the same makespans that are guaranteed to be minimal by the optimality of the A* search since their heuristic functions are admissible. In addition, the presented symbolic A* search dramatically outperforms the explicit-state A* in terms of the computational cost and can handle some larger cases that cannot be solved by the explicit-state method.

The third example is a more complex system adapted from Chen *et al.* (2011). It has 12 resources R_1–R_{12}, five input buffers I_1–I_5, and five output buffers O_1–O_5.

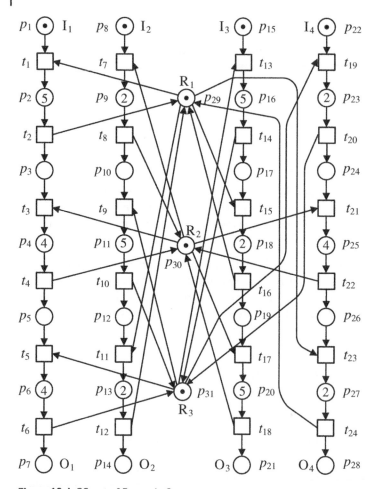

Figure 12.4 SC-net of Example 2.

Five types of parts, P_1–P_5, are processed in the system. The processing routes are as follows:

$$P_1: I_1 \rightarrow R_1 \rightarrow R_6 \rightarrow R_2 \rightarrow R_7 \rightarrow R_3 \rightarrow O_1$$
$$\text{or } I_1 \rightarrow R_1 \rightarrow R_8 \rightarrow R_3 \rightarrow R_9 \rightarrow R_3 \rightarrow O_1$$
$$P_2: I_2 \rightarrow R_2 \rightarrow R_{11} \rightarrow R_2 \rightarrow R_7 \rightarrow R_5 \rightarrow O_2$$
$$P_3: I_3 \rightarrow R_3 \rightarrow R_9 \rightarrow R_3 \rightarrow R_8 \rightarrow R_1 \rightarrow O_3$$
$$P_4: I_4 \rightarrow R_4 \rightarrow R_{11} \rightarrow R_2 \rightarrow R_{10} \rightarrow R_2 \rightarrow O_4$$
$$\text{or } I_4 \rightarrow R_4 \rightarrow R_9 \rightarrow R_3 \rightarrow R_7 \rightarrow R_2 \rightarrow O_4$$
$$P_5: I_5 \rightarrow R_5 \rightarrow R_{12} \rightarrow R_1 \rightarrow R_6 \rightarrow R_1 \rightarrow O_5$$

Table 12.4 Scheduling results for Example 2.

p_1	p_8	p_{15}	p_{22}	Makespan	Explicit A*($h=0$)		Symbolic A*($h=0$)			Symbolic A*($h=h_{BDD}$)		
					N_E	T(s)	N_E	N_{BDD}	T(s)	N_E	N_{BDD}	T(s)
1	1	1	1	16	2 696	1.09	1.75×10^4	2.36×10^5	4.71	2.68×10^3	5.29×10^4	1.65
2	1	1	1	20	13 934	35.54	1.51×10^5	1.15×10^6	28.13	8.18×10^3	1.14×10^5	2.99
2	2	1	1	25	71 673	1041.89	1.05×10^6	4.35×10^6	146.86	6.18×10^4	4.47×10^5	17.97
2	2	2	1	30	—	—	5.30×10^6	1.28×10^7	798.84	3.78×10^5	1.37×10^6	91.17
2	2	2	2	32	—	—	2.13×10^7	2.88×10^7	3442.18	1.40×10^6	3.12×10^6	379.28

Table 12.5 Operation times of Example 3.

Operation	Operation time	Operation	Operation time
$O_{1,1,1}$	5	$O_{3,4,8}$	2
$O_{1,2,6}$	2	$O_{3,5,1}$	7
$O_{1,2,8}$	3	$O_{4,1,4}$	5
$O_{1,3,2}$	4	$O_{4,2,11}$	3
$O_{1,3,3}$	2	$O_{4,2,9}$	2
$O_{1,4,7}$	6	$O_{4,3,2}$	4
$O_{1,4,9}$	1	$O_{4,3,3}$	4
$O_{1,5,3}$	2	$O_{4,4,10}$	1
$O_{2,1,2}$	3	$O_{4,4,7}$	3
$O_{2,2,11}$	2	$O_{4,5,2}$	2
$O_{2,3,2}$	3	$O_{5,1,5}$	6
$O_{2,4,7}$	2	$O_{5,2,12}$	3
$O_{2,5,5}$	4	$O_{5,3,1}$	3
$O_{3,1,3}$	5	$O_{5,4,6}$	1
$O_{3,2,9}$	2	$O_{5,5,1}$	7
$O_{3,3,3}$	3		

The operation time of each activity is given in Table 12.5. Its SC-net model is shown in Figure 12.5 that has 38 transitions and 53 places in which $P_S = \{p_1, p_{10}, p_{16}, p_{22}, p_{31}\}$, $P_E = \{p_{49} - p_{53}\}$, and $P_R = \{p_{37} - p_{48}\}$, while the rest are activity places. The symbolic A* search with h_{BDD} is used to solve it. Table 12.6 shows the obtained schedule in the form of a sequence of transition firings, which represent a series of actions of the system. Its makespan is 36 that is guaranteed to be minimal since h_{BDD} is admissible. The number of expanded states is 9.96×10^5, the number of BDD nodes for these states is 1.46×10^7, and the computational time is about 41 minutes. In contrast, the explicit-state A* with any admissible heuristic function in Huang *et al.* (2014), Lee and Lee (2010), Luo *et al.* (2015), Mejía (2002), Moro *et al.* (2002b), and Xiong and Zhou (1998) cannot solve it in an acceptable period of time (1 day).

12.6 Concluding Remarks

This chapter presents a symbolic A* method that uses succinct data structures of BDDs to reason about sets of states and executes an efficient functional exploration in the reachability graphs of place-timed SC-nets. The experiments show that the

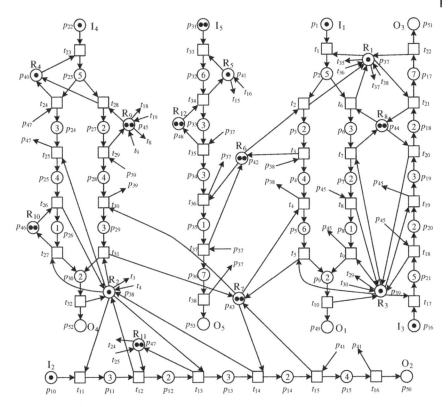

Figure 12.5 SC-net of Example 3.

method is much more efficient than the existing explicit-state A* searches in reachability graphs. So, it is a promising way to alleviate the state explosion problem in RAS scheduling.

In the future, the presented method can be extended in several aspects. For instance, it is challenging to develop more informed heuristics for the symbolic A* search within the reachability graphs of SC-nets. Refining the symbolic A* method to efficiently schedule other discrete event systems, such as steel production, semiconductor manufacturing, demanufacturing, and cloud task allocation (Bi *et al.*, 2017; Fu *et al.*, 2019; Guo *et al.*, 2016, 2018; Han *et al.*, 2017; Hu *et al.*, 2013; Li *et al.*, 2018; Pan *et al.*, 2018; Qiao *et al.*, 2020; Tan *et al.*, 2019; Tang *et al.*, 2001; Yuan *et al.*, 2017, 2018; Zhang & Zhou, 2018) and dealing with multicriteria decision-making problems (Liu *et al.*, 2019; Sangiorgio *et al.*, 2017) are other interesting issues. The introduction of recent AI technologies (Dong & Zhou, 2017; Gao & Zhou, 2019; Ghahramani *et al.*, 2020; Li *et al.*, 2015) into such A* methods is worth investigating.

Table 12.6 The obtained optimal schedule for Example 3 (makespan = 36).

Transition	Fire time	Transition	Fire time	Transition	Fire time
t_{33}	0	t_{20}	10	t_3	22
t_{23}	0	t_{36}	12	t_{36}	22
t_{17}	0	t_{34}	12	t_{37}	22
t_1	0	t_{26}	12	t_4	26
t_{24}	5	t_{21}	12	t_{38}	29
t_{18}	5	t_{11}	12	t_{37}	29
t_2	5	t_{12}	15	t_{27}	29
t_{34}	6	t_{13}	17	t_5	32
t_{33}	6	t_{22}	19	t_{38}	36
t_{19}	7	t_{35}	19	t_{32}	36
t_{25}	8	t_{14}	20	t_{16}	36
t_{35}	9	t_{15}	22	t_{10}	36

12.7 Bibliographical Notes

The chapter presents an efficient symbolic method to quickly and optimally schedule timed SC-nets for RASs based on BDDs. For more applications of timed Petri nets, please refer to the book (Wang, 1998). A comprehensive review of the research on Petri net-based scheduling can be seen in Tuncel and Bayhan (2007). For more details about the heuristic search algorithms, please refer to Edelkamp and Schroedl (2012) and Pearl (1984). The symbolic methods that efficiently compute an untimed Petri net's structures (such as reachable markings, legal markings, and minimal siphons) can be found in Chen *et al.* (2011), Chen and Liu (2013), and Pastor *et al.* (2001). The material presented in this chapter for the symbolic scheduling based on timed Petri nets mainly comes from our work (Huang *et al.*, 2018*b*). The implementation codes of the presented method and the experimental files of all examples in this chapter are available in Huang (2019*b*) for the reader's examination and reference.

13

Open Problems

This chapter provides several interesting open problems in the development of deadlock prevention and system scheduling methods based on Petri nets such as structural analysis of generalized nets, robust supervisor synthesis with unreliable resources, strategies to alleviate the state explosion problem, optimization of symbolic variable ordering, multiobjective heuristic search, and parallel heuristic search. They facilitate readers in finding suitable topics for their future research.

13.1 Structural Analysis of Generalized Nets

In Petri net models, deadlocks can be characterized by strict minimal siphons (SMSs) or maximal perfect resource transition circuits (MPRT-circuits) (Feng *et al.*, 2020; Xing *et al.*, 2011). So, structural analysis methods are usually employed to design Petri net supervisors that avoid emptied siphons or saturated MPRT-circuits for deadlock prevention.

However, in the worst case, both the numbers of siphons and MPRT-circuits exponentially grow with respect to the net size. Thus, how to efficiently compute them for large-scale Petri nets is a challenging issue. Furthermore, the sufficient and necessary conditions for deadlock control with siphons in generalized Petri nets, such as S^4R nets and DP-nets, have not been developed. Finally, how to design behavior-maximal supervisors for generalized Petri nets based on the structural analysis remains widely open.

13.2 Robust Supervisor Synthesis with Unreliable Resources

Most studies in the literature assume that resources do not fail. Actually, resource failures are inevitable in practical resource allocation systems (RASs) and they

Supervisory Control and Scheduling of Resource Allocation Systems: Reachability Graph Perspective, First Edition. Bo Huang and MengChu Zhou.

may cause deadlocks. Thus, developing a robust deadlock control policy to ensure that deadlocks cannot occur in the RASs whose resources may break down is also important.

To address it, some researchers have focused on robust deadlock analysis and avoidance in such systems. For example, Reveliotis (1999) considers a scenario in which objects requiring a failed resource can be rerouted or removed by using human intervention. Park and Lim (1999) deal with the existence problem of robust supervisors. Some supervisor synthesis methods are proposed in Chew and Lawley (2006), Chew *et al.* (2009) and Lawley and Sulistyono (2002) for unstable systems based on the banker's algorithm and central buffer constraints. There also exist some methods to determine the feasibility of production with a set of resource failures that are modeled as the extraction of tokens from different classes of Petri nets, including controlled production Petri nets (Hsieh, 2003), controlled assembly Petri nets (Hsieh, 2004), controlled assembly/disassembly Petri nets (Hsieh, 2006), controlled assembly Petri nets with alternative routes (Hsieh, 2007), collaborative Petri nets (Hsieh, 2008), and non-ordinary controlled flexible assembly Petri nets with uncertainties (Hsieh, 2010). In recent years, some robust deadlock prevention policies are also proposed for the Petri nets of RASs with unreliable resources (Feng *et al.*, 2017; Liu, 2014; Luo *et al.*, 2017; Wu *et al.*, 2018), which have the following properties: i) they can prevent deadlocks for a plant net when all resources work normally, ii) deadlocks are prevented even if some resources fail and are removed for a repair at any time, and iii) deadlocks disappear after the repaired resources are returned. However, these methods mainly focus on single-unit RASs and systems with a single unreliable resource. How to robustly handle the deadlocks for more complex RASs such as disjunctive/conjunctive RASs (i.e., AND/OR RASs) with multiple unreliable resources is not an easy task. A recent survey about this topic is from Du *et al.* (2020).

13.3 Alleviation of the State Explosion Problem

Reachability analysis methods are an important approach to analyze and verify the behavior of RASs. These methods need to enumerate and analyze the reachable states of the underlying systems. Unfortunately, the set of reachable states of RASs is often far larger than those that can be handled in a computer due to the state explosion problem, i.e., the number of reachable states increases exponentially with respect to the net size. In the literature, many approaches have been proposed to alleviate the problem.

The first kind of methods is to apply suitable transformations to Petri net models before the construction of their reachability graphs. For example, a net reduction method is given in Uzam (2004) and Yu *et al.* (2018) that uses the net reduction method to simplify the models to make calculations easier. The second kind of methods belongs to the symmetry approach. Many systems exhibit symmetry in one form or another. For instance, a system may contain several identical components that are coupled to each other and the rest of the system in a regular way. The basic idea of the state space reduction with symmetry is to store only one state from a set of equivalent states (Jensen & Kristensen, 2009). The third kind of methods is to use binary decision diagrams (BDDs) to compactly represent sets of states and efficiently perform set operations such as union, intersection, complementation, and equivalence. Chapter 12 has used it to efficiently schedule SC-nets for RASs. Except the above mentioned ones, the novel use of ordinary differential equations for deadlock analysis of large size Petri nets is pioneered in Ding *et al.* (2014) and its analysis is of polynomial complexity only.

The above approaches can alleviate the state explosion problem in reachability analysis methods, while none of today's techniques can solve the problem once and for all. Also, it seems unlikely that a perfect solution would be found in the near future. However, it is interesting to combine multiple methods, such as the symmetry approach and ordered BDDs, to further alleviate the problem.

13.4 Optimization of Symbolic Variable Ordering

Resource allocation systems belong to highly concurrent systems that suffer from the state explosion problem due to an exponential increase in the number of reachable states as system size and/or initial marking increases. The problem can be alleviated by using symbolic methods such as BDDs to represent large sets of states with small data structures. When compared with the explicit-state search that operates on individual states, the symbolic search efficiently handles sets of states in each operation.

In the symbolic methods, the Boolean variable ordering is a key factor for the efficiency of such methods since it largely influences the size of used BDDs, varying from linearly to exponentially. However, the variable ordering problem is co-NP-complete (Bryant, 1986). It is thus practically infeasible to design an optimal method to handle it. To find an optimal or good variable ordering, there are many approaches that can be classified into static variable ordering and dynamic variable reordering. In static variable ordering, dependent variables are usually brought close together in order to reduce BDD sizes, typically with causal graphs. However, Kissmann and Hoffmann (2014) show that even for simple search problems, the variable orders obtained by causal-graph-based methods may be exponentially worse than the optimal order. On the other hand, dynamic variable reordering is a much more efficient method that can be used to reduce BDD size, but it is very time-consuming (Kissmann & Hoffmann, 2014). Therefore, how to design an effective variable ordering method suitable for the symbolic search within reachability graphs is an interesting issue.

13.5 Multiobjective Scheduling

For the single-objective scheduling of RASs based on Petri nets, an A*-based search and its improvements within reachability graphs can reduce computational requirements with heuristic information and help obtain optimal or near-optimal schedules. But in many practical RASs, there exist multiple, conflicting, and non-commensurate objectives such as the minimizations of makespan, tardiness penalty, and energy cost for the system scheduling (Gao *et al.*, 2019a, 2019b; Hou *et al.*, 2017; Zuo *et al.*, 2015). Thus, searching for optimal schedules with several conflicting objectives is important to schedule such systems.

Different from the single-objective searches within reachability graphs, multiobjective searches have more than one attribute for each activity place, which introduces a partial order relation among different paths in the reachability graphs. How to evolve such multiattribute nets and design heuristic functions suitable for the multiobjective search of these nets are critical. On the other hand, for the multiobjective optimization, it is usually impossible for all objectives to be simultaneously satisfied. Instead, each objective can only be satisfied as much as possible. Such solutions are called non-dominated or Pareto-optimal solutions (Huang *et al.*, 2020; Kang *et al.*, 2019), because they cannot satisfy one objective without sacrificing another. Without any further information, a non-dominated solution cannot be said to be better than any other non-dominated solution. Thus, the user needs to find as many non-dominated solutions as possible. However, the number of non-dominated solutions often grows exponentially with the size of the underlying problem. To accelerate the multiobjective search, a simple strategy is to convert a multiobjective search into a single-objective one by using a weighting method that combines several criteria into a single scalar value (Kang *et al.*, 2017). When such a method is used to find multiple solutions, it has to be executed many times, hopefully finding a different solution with each run. In addition, it is difficult to give an accurate weight to each criterion, which makes the quality of the obtained solutions uncontrollable. Therefore, how to speed up the multi-objective search within reachability graphs for RASs while the solutions' quality is still optimal or controllable interests us.

13.6 Anytime Heuristic Scheduling

An optimal search is usually infeasible for real-world scheduling problems due to the time limitation and the NP-hard nature of scheduling problems. In these cases, anytime search algorithms (Baruwa *et al.*, 2015; Hansen & Zhou, 2007; Likhachev *et al.*, 2005) are useful because they usually find a schedule very quickly (although it may be of limited quality) and then continually work on improving the solution until the allocated time expires. They are suitable for the system scheduling with varying or uncertain time constraints since they can usually have a solution whenever they are stopped and the quality of the solution can be continuously improved with additional time.

A common feature of anytime search algorithms is the ability to prune some nodes in the OPEN list after the first schedule is obtained. For example, the anytime weighted A* (Hansen & Zhou, 2007) reduces the size of OPEN by using a weighted heuristic function. It uses two evaluation functions, i.e., $f(n)$ that is the same as that in A* and $f'(n) = g(n) + h'(n)$ in which $h'(n)$ is an inadmissible heuristic that is computed as $h'(n) = \omega \times h(n)$ where the parameter $\omega \geq 1$ adjusts the tradeoff between computational time and results' quality. The algorithm uses $f'(n)$ to select a node with the smallest f'-value from the OPEN list. Then, the algorithm checks if the f-value of the node is greater than the smallest f-value found so far. If yes, the node is pruned; otherwise, the node is expanded. The termination condition of the algorithm can also be based on the f-values and that is why it converges to an optimal schedule if time permits.

We see that weighted A* searches play an important role in anytime A* algorithms. Chapter 9 presents two weighted A* search algorithms. An interesting issue is how to incorporate them into anytime A* algorithms to efficiently find schedules within the reachability graphs of Petri nets.

13.7 Parallel Heuristic Search

Parallel searches are designed for the pathfinding problem that simultaneously uses many processing devices (such as processors, processor cores, and computing nodes) to accelerate search speed. They explore the possibilities of concurrent execution to make the search faster. In terms of memory architecture, they can be classified into shared memory algorithms (e.g., with multicore processors) and distributed memory algorithms (e.g., with workstation clusters). The general algorithmic problems in parallel heuristic searches include compatibility with a machine architecture, choice of shared data structures, and compromise between processing and communication overhead (Burns *et al.*, 2010; Edelkamp & Schroedl, 2012; Zaitsev *et al.*, 2019). In addition, the acceleration of a parallel search compared to a serial search depends on the properties of the application at hand.

We know that a heuristic A* algorithm maintains two lists of nodes, i.e., OPEN and CLOSED. The OPEN list is a search frontier whose nodes have been generated but not yet expanded and are sorted by their f-values. The OPEN list is typically implemented by using a priority queue. The CLOSED list contains all expanded nodes, allowing the search to detect states that can be reached via multiple paths in the search space and avoid expanding them repeatedly. The CLOSED list is often implemented as a hash table. The central challenge in the implementation of a parallel A* search is to avoid contention among threads when accessing the OPEN and CLOSED lists, for example, how to efficiently and concurrently expand nodes in OPEN while guaranteeing the quality of the results and how to efficiently perform parallel duplicate detection in CLOSED.

13.8 Bidirectional Heuristic Search

A bidirectional search is a distributed search method starting from both the initial and the goal node. A bidirectional breadth-first search has a low price. However, it is not true for a bidirectional A* search. The inefficiency of the bidirectional A* search is not due to the problem of search frontiers passing each other. In fact, its search frontiers usually go through each other. When two frontiers meet, even if the two found paths are optimal, the concatenation is not always optimal. Thus, it needs a complex termination condition, i.e., the cost of the best solution found so far should not be greater than the maximum of the two minimal f-values in either of the Open lists, to ensure the optimality of the obtained schedule. Therefore, the main reason for the inefficiency of the bidirectional A* is that its major search effort is spent on finding the solutions that are better than the ones already found and on proving that there is indeed no other better solution.

Therefore, it is challenging to design a bidirectional A* search that can reduce the computational burden regarding the number of searched nodes. In López and Junghanns (2002), a perimeter search that is a variant of the bidirectional A* search is proven to be a promising method to enhance the search efficiency. Thus, how to apply it to the path-finding within reachability graphs is an interesting issue. Furthermore, the combination of bidirectional breadth-first search with the symbolic representations presented in Chapter 12 to further speed up the search is another important topic.

13.9 Computing and Scheduling with GPUs

A graphics processing unit (GPU) is a general-purpose, multi-threaded, and parallel co-processor that has hundreds of cores compared to the small number of cores in a CPU. GPUs are originally dedicated to graphics processing in which the same calculations have to be performed on thousands of image points. Thus, engineers have assembled hundreds of cores in a GPU chip. The GPU architecture mimics a single-instruction-multiple-data computer that runs the same instructions on all processors. In recent years, the use of GPUs as general-purpose computing devices has risen significantly, speeding up many non-graphics programs that exhibit lots of parallelism with low synchronization requirements (Bach *et al.*, 2019; Hwu, 2011; Zhou *et al.*, 2017). However, GPU computing for analyzing and controlling Petri net models has never been investigated in the literature. The Petri nets' evolution and analysis are usually based on matrix and vector operations, which can enjoy the efficient implementation on GPUs. Therefore, it is promising to utilize GPUs to achieve efficiency in the deadlock prevention and system scheduling of RASs based on Petri nets.

References

Abdallah, I. B., Hoda A. Elmaraghy, & Tarek Elmekkawy (2002). 'Deadlock-free scheduling Petri nets'. *International Journal of Production Research* 40(12), 2733–2756.

Azaiez, M. N., & S. Shaza Al Sharif (2005). 'A 0–1 goal programming model for nurse scheduling'. *Computers & Operations Research* 32(3), 491–507.

Bach, L., Geir Hasle, & Christian Schulz (2019). 'Adaptive large neighborhood search on the graphics processing unit'. *European Journal of Operational Research* 275(1), 53–66.

Bai, L., Naiqi Wu, Zhiwu Li, & MengChu Zhou (2016). 'Optimal one-wafer cyclic scheduling and buffer space configuration for single-arm multicluster tools with linear topology.' *IEEE Transactions on Systems Man, and Cybernetics: Systems* 46(10), 1456–1467.

Banaszak, Z. A., & Bruce H. Krogh (1990). 'Deadlock avoidance in flexible manufacturing systems with concurrently competing process flows'. *IEEE Transactions on Robotics and Automation* 6(6), 724–734.

Barkaoui, K., & I. Ben Abdallah (1996). Analysis of a resource allocation problem in FMS using structure theory of Petri nets. In '1st International Workshop on Manufacturing and Petri Nets'. Osaka, Japan. pp. 1–15.

Baruwa, O. T., Miquel Angel Piera, & Antoni Guasch (2015). 'Deadlock-free scheduling method for flexible manufacturing systems based on timed colored Petri nets and anytime heuristic search'. *IEEE Transactions on Systems, Man, and Cybernetics: Systems* 45(5), 831–846.

Basile, F., Roberto Cordone, & Luigi Piroddi (2013). 'Integrated design of optimal supervisors for the enforcement of static and behavioral specifications in Petri net models'. *Automatica* 49(11), 3432–3439.

Beaudouin-Lafon, M., Wendy E. Mackay, Peter Andersen, Paul Janecek, Mads Jensen, Michael Lassen, Kasper Lund, Kjeld Mortensen, Stephanie Munck, Anne Ratzer et al. (2001). CPN/tools: A post-WIMP interface for editing and simulating coloured Petri nets. In 'International Conference on Application and Theory of Petri Nets'. Springer, Newcastle, UK. pp. 71–80.

Supervisory Control and Scheduling of Resource Allocation Systems: Reachability Graph Perspective,
First Edition. Bo Huang and MengChu Zhou.
© 2020 The Institute of Electrical and Electronics Engineers, Inc. Published 2020 by John Wiley & Sons, Inc.

Berthomieu, B., P.-O. Ribet, & François Vernadat (2004). 'The tool TINA–construction of abstract state spaces for Petri nets and time Petri nets'. *International Journal of Production Research* 42(14), 2741–2756.

Bryant, R. (1986). 'Graph-based algorithms for boolean function manipulation'. *IEEE Transactions on Computers* 35(8), 677–691.

Burkard, R. E., & Johannes Hatzl (2005). 'Review, extensions and computational comparison of MILP formulations for scheduling of batch processes'. *Computers & Chemical Engineering* 29(8), 1752–1769.

Burns, E., Sofia Lemons, Wheeler Ruml, & Rong Zhou (2010). 'Best-first heuristic search for multicore machines'. *Journal of Artificial Intelligence Research* 39, 689–743.

Cao, Z., C. Lin, M. Zhou, & R. Huang (2019). 'Scheduling semiconductor testing facility by using cuckoo search algorithm with reinforcement learning and surrogate modeling'. *IEEE Transactions on Automation Science and Engineering* 16(2), 825–837.

Chen, H., N. Wu, & MengChu Zhou (2016). 'A novel method for deadlock prevention of AMS by using resource-oriented Petri nets'. *Information Sciences* 363, 178–189.

Chen, P.-H., & Seyed Mohsen Shahandashti (2009). 'Hybrid of genetic algorithm and simulated annealing for multiple project scheduling with multiple resource constraints'. *Automation in Construction* 18(4), 434–443.

Chen, Y., & Gaiyun Liu (2013). 'Computation of minimal siphons in Petri nets by using binary decision diagrams'. *ACM Transactions on Embedded Computing Systems* 12(1), 3.

Chen, Y., & Kamel Barkaoui (2014). 'Maximally permissive Petri net supervisors for flexible manufacturing systems with uncontrollable and unobservable transitions'. *Asian Journal of Control* 16(6), 1646–1658.

Chen, Y., & ZhiWu Li (2011). 'Design of a maximally permissive liveness-enforcing supervisor with a compressed supervisory structure for flexible manufacturing systems'. *Automatica* 47(5), 1028–1034.

Chen, Y., & ZhiWu Li (2012). 'On structural minimality of optimal supervisors for flexible manufacturing systems'. *Automatica* 48(10), 2647–2656.

Chen, Y., & ZhiWu Li (2013). *Optimal Supervisory Control of Automated Manufacturing Systems*. CRC Press, Boca Raton, FL, USA.

Chen, Y., ZhiWu Li, & Kamel Barkaoui (2014a). 'Maximally permissive liveness-enforcing supervisor with lowest implementation cost for flexible manufacturing systems'. *Information Sciences* 256, 74–90.

Chen, Y., ZhiWu Li, & MengChu Zhou (2012). 'Behaviorally optimal and structurally simple liveness-enforcing supervisors of flexible manufacturing systems'. *IEEE Transactions on Systems, Man, and Cybernetics—Part A: Systems and Humans* 42(3), 615–629.

Chen, Y., ZhiWu Li, Kamel Barkaoui, & Murat Uzam (2014b). 'New Petri net structure and its application to optimal supervisory control: Interval inhibitor arcs'. *IEEE Transactions on Systems, Man, and Cybernetics: Systems* 44(10), 1384–1400.

Chen, Y., ZhiWu Li, Kamel Barkaoui, NaiQi Wu, & MengChu Zhou (2017). 'Compact supervisory control of discrete event systems by Petri nets with data inhibitor arcs'. *IEEE Transactions on Systems, Man, and Cybernetics: Systems* 47(2), 364–379.

Chen, Y., ZhiWu Li, Mohamed Khalgui, & Olfa Mosbahi (2011). 'Design of a maximally permissive liveness-enforcing Petri net supervisor for flexible manufacturing systems'. *IEEE Transactions on Automation Science and Engineering* 8(2), 374–393.

Chew, S. F., & Mark A. Lawley (2006). 'Robust supervisory control for production systems with multiple resource failures'. *IEEE Transactions on Automation Science and Engineering* 3(3), 309–323.

Chew, S. F., Shengyong Wang, & Mark A. Lawley (2009). 'Robust supervisory control for product routings with multiple unreliable resources'. *IEEE Transactions on Automation Science and Engineering* 6(1), 195–200.

Coffman, E. G., Melanie Elphick, & Arie Shoshani (1971). 'System deadlocks'. *ACM Computing Surveys* 3(2), 67–78.

Cordone, R., & Luigi Piroddi (2011). Monitor optimization in Petri net control. In IEEE International Conference on Automation Science and Engineering. IEEE, Trieste, Italy. pp. 413–418.

Cordone, R., & Luigi Piroddi (2013). 'Parsimonious monitor control of Petri net models of flexible manufacturing systems'. *IEEE Transactions on Systems, Man, and Cybernetics: Systems* 43(1), 215–221.

Cordone, R., Luca Ferrarini, & Luigi Piroddi (2005). 'Enumeration algorithms for minimal siphons in Petri nets based on place constraints'. *IEEE Transactions on Systems, Man, and Cybernetics—Part A: Systems and Humans* 35(6), 844–854.

Ding, Z., Changjun Jiang, & MengChu Zhou (2013). 'Design, analysis and verification of real-time systems based on time Petri net refinement'. *ACM Transactions in Embedded Computing Systems* 12(1), 4:1–18.

Ding, Z., MengChu Zhou, & Shouguang Wang (2014). 'Ordinary differential equation based deadlock detection.' *IEEE Transactions on Systems, Man, and Cybernetics: Systems* 44(10), 1435–1454.

Dominic, P. D., Sathya Kaliyamoorthy, & M. Saravana Kumar (2004). 'Efficient dispatching rules for dynamic job shop scheduling'. *International Journal of Advanced Manufacturing Technology* 24(1–2), 70–75.

Dong, W., & MengChu Zhou (2017). 'A supervised learning and control method to improve particle swarm optimization algorithms.' *IEEE Transactions on Systems, Man, and Cybernetics: Systems* 47(7), 1149–1159.

Du, N., Hesuan Hu, & MengChu Zhou (2020). 'A survey on robust deadlock control policies for automated manufacturing systems with unreliable resources.' *IEEE Transactions on Automation Science and Engineering* 17(1), 389–406.

Du, Y., Wei Tan, & MengChu Zhou (2014). 'Timed compatibility analysis of web service composition: a modular approach based on Petri nets.' *IEEE Transactions on Automation Science and Engineering* 11(2), 594–606.

Edelkamp, S., & Stefan Schroedl (2012). *Heuristic Search: Theory and Applications.* Elsevier, Waltham, MA, USA.

Elmekkawy, T. Y., & Hoda A. Elmaraghy (2003). 'Efficient search of Petri nets for deadlock-free scheduling in FMSs using heuristic functions'. *International Journal of Computer Integrated Manufacturing* 16(1), 14–24.

Ezpeleta, J., & Laura Recalde (2004). 'A deadlock avoidance approach for nonsequential resource allocation systems'. *IEEE Transactions on Systems, Man, and Cybernetics—Part A: Systems and Humans* 34(1), 93–101.

Ezpeleta, J., Fernando García-Vallés, & José Manuel Colom (1998). A class of well structured Petri nets for flexible manufacturing systems. In 'Application and Theory of Petri Nets 1998'. Springer, Lisbon, Portugal. pp. 64–83.

Ezpeleta, J., Fernando Tricas, Fernando Garcia-Valles, & José Manuel Colom (2002). 'A banker's solution for deadlock avoidance in FMS with flexible routing and multiresource states'. *IEEE Transactions on Robotics and Automation* 18(4), 621–625.

Ezpeleta, J., Jose Manuel Colom, & Javier Martinez (1995). 'A Petri net based deadlock prevention policy for flexible manufacturing systems'. *IEEE Transactions on Robotics and Automation* 11(2), 173–184.

Fanti, M. P., & MengChu Zhou (2004). 'Deadlock control methods in automated manufacturing systems'. *IEEE Transactions on Systems, Man, and Cybernetics—Part A: Systems and Humans* 34(1), 5–22.

Fanti, M. P., Alessandro Giua, & Carla Seatzu (2006). 'Monitor design for colored Petri nets: An application to deadlock prevention in railway networks'. *Control Engineering Practice* 14(10), 1231–1247.

Fanti, M. P., Bruno Maione, Saverio Mascolo, & A. Turchiano (1997). 'Event-based feedback control for deadlock avoidance in flexible production systems'. *IEEE Transactions on Robotics and Automation* 13(3), 347–363.

Fanti, M. P., Guido Maione, & Biagio Turchiano (2002). 'Design of supervisors to avoid deadlock in flexible assembly systems'. *International Journal of Flexible Manufacturing Systems* 14(2), 153–171.

Feng, Y., Keyi Xing, Zhenxin Gao, & Yunchao Wu (2017). 'Transition cover-based robust Petri net controllers for automated manufacturing systems with a type of unreliable resources'. *IEEE Transactions on Systems, Man, and Cybernetics: Systems* 47(11), 3019–3029.

Feng, Y., Keyi Xing, Mengchu Zhou, & Huixia Liu (2020). 'Liveness analysis and deadlock control for automated manufacturing systems with multiple resource requirements.' *IEEE Transactions on Systems, Man, and Cybernetics: Systems* 50(2), 525–538.

Feng, Y., MengChu Zhou, Guangdong Tian, Zhiwu Li, Zhifeng Zhang, Qin Zhang, & Jianrong Tan (2019). 'Target disassembly sequencing and scheme evaluation for CNC machine tools using improved multiobjective ant colony algorithm and fuzzy integral.' *IEEE Transactions on Systems, Man, and Cybernetics: Systems* 49(12), 2438–2451.

Fu, Y., Mengchu Zhou, Xiwang Guo, & Liang Qi (2019). 'Artificial-molecule-based chemical reaction optimization for flow shop scheduling problem with deteriorating and learning effects'. *IEEE Access* 7, 53429–53440.

Gao, K., Fajun Yang, MengChu Zhou, & Quanke Pan (2019a). 'Flexible job-shop rescheduling for new job insertion by using discrete Jaya algorithm'. *IEEE Transactions on Cybernetics* 49(5), 1944–1955.

Gao, S., MengChu Zhou, Yirui Wang, Jiujun Cheng, Hanaki Yachi, & Jiahai Wang (2019b). 'Dendritic neuron model with effective learning algorithms for classification, approximation and prediction'. *IEEE Transactions on Neural Networks and Learning Systems* 30(2), 601–614.

Ghaffari, A., Nidhal Rezg, & Xiaolan Xie (2003). 'Design of a live and maximally permissive Petri net controller using the theory of regions'. *IEEE Transactions on Robotics and Automation* 19(1), 137–141.

Ghahramani, M., Yan Qiao, MengChu Zhou, Adrian O. Hagan, & James Sweeney (2020). 'AI-based modeling and data-driven evaluation for smart manufacturing processes'. *IEEE/CAA Journal of Automatica Sinica* 7(4), 948–959.

Giua, A., Frank Dicesare, & Manuel Silva (1992). Generalized mutual exclusion constraints on nets with uncontrollable transitions. In 'IEEE International Conference on Systems, Man, and Cybernetics'. Chicago, IL, USA. pp. 974–979.

Guo, X., Shixin Liu, MengChu Zhou, & Guangdong Tian (2016). 'Disassembly sequence optimization for large-scale products with multi-resource constraints using scatter search and Petri nets'. *IEEE Transactions on Cybernetics* 46(11), 2435–2446.

Guo, X., Shixin Liu, MengChu Zhou, & Guangdong Tian (2018). 'Dual-objective program and scatter search for the optimization of disassembly sequences subject to multiresource constraints'. *IEEE Transactions on Automation Science and Engineering* 15(3), 1091–1103.

Han, Z., Jun Zhao, & Wei Wang (2017). 'An optimized oxygen system scheduling with electricity cost consideration in steel industry'. *IEEE/CAA Journal of Automatica Sinica* 4(2), 216–222.

Hansen, E. A., & Rong Zhou (2007). 'Anytime heuristic search'. *Journal of Artificial Intelligence Research* 28, 267–297.

Hou, Y., NaiQi Wu, MengChu Zhou, & ZhiWu Li (2017). 'Pareto-optimization for scheduling of crude oil operations in refinery via genetic algorithm'. *IEEE Transactions on Systems, Man, and Cybernetics: Systems* 47(3), 517–530.

Hou, Y., ZhiWu Li, Abdulrahman M. Al-Ahmari, Abdul-Aziz Mohammed El-Tamimi, & Emad Abouel Nasr (2014). 'Extended elementary siphons and their application to liveness-enforcement of generalized Petri nets'. *Asian Journal of Control* 16(6), 1789–1810.

Hrúz, B., & MengChu Zhou (2007). *Modeling and Control of Discrete-Event Dynamic Systems: With Petri Nets and Other Tools*. Springer, London, UK.

Hsieh, F.-S. (2003). 'Robustness of deadlock avoidance algorithms for sequential processes'. *Automatica* 39(10), 1695–1706.

Hsieh, F.-S. (2004). 'Fault-tolerant deadlock avoidance algorithm for assembly processes'. *IEEE Transactions on Systems, Man, and Cybernetics—Part A: Systems and Humans* 34(1), 65–79.

Hsieh, F.-S. (2006). 'Robustness analysis of Petri nets for assembly/disassembly processes with unreliable resources'. *Automatica* 42(7), 1159–1166.

Hsieh, F.-S. (2007). 'Analysis of flexible assembly processes based on structural decomposition of Petri nets'. *IEEE Transactions on Systems, Man, and Cybernetics—Part A: Systems and Humans* 37(5), 792–803.

Hsieh, F.-S. (2008). 'Robustness analysis of holonic assembly/disassembly processes with Petri nets'. *Automatica* 44(10), 2538–2548.

Hsieh, F.-S. (2010). 'Robustness analysis of non-ordinary Petri nets for flexible assembly systems'. *International Journal of Control* 83(5), 928–939.

Hu, H., MengChu Zhou, & Zhiwu Li (2010). 'Low-cost and high-performance supervision in ratio-enforced automated manufacturing systems using timed Petri nets'. *IEEE Transactions on Automation Science and Engineering* 7(4), 933–944.

Hu, H., MengChu Zhou, Zhiwu Li, & Ying Tang (2013). 'Deadlock-free control of automated manufacturing systems with flexible routes and assembly operations using Petri nets'. *IEEE Transactions on Industrial Informatics* 9(1), 109–121.

Huang, B. (2019a). 'Source codes and experimental files'. http://github.com/huang-njust/schedule. Accessed: May 19, 2019.

Huang, B. (2019b). 'Source codes and experimental files'. http://github.com/huang-njust/bdd. Accessed: May 19, 2019.

Huang, B., & Yamin Sun (2005). 'Improved methods for scheduling flexible manufacturing systems based on Petri nets and heuristic search'. *Journal of Control Theory and Applications* 3(2), 139–144.

Huang, B., Hang Zhu, GongXuan Zhang, & XianLing Lu (2015a). 'On further reduction of constraints in nonpure Petri net supervisors for optimal deadlock control of flexible manufacturing systems'. *IEEE Transactions on Systems, Man, and Cybernetics: Systems* 45(3), 542–543.

Huang, B., MengChu Zhou, & GongXuan Zhang (2015b). 'Synthesis of Petri net supervisors for FMS via redundant constraint elimination'. *Automatica* 61, 156–163.

Huang, B., MengChu Zhou, GongXuan Zhang, Ahmed Chiheb Ammari, Ahmed Alabdulwahab, & Ayman G. Fayoumi (2015c). 'Lexicographic multiobjective interger programming for optimal and structurally minimal Petri net supervisors of automated manufacturing systems'. *IEEE Transactions on Systems, Man, and Cybernetics: Systems* 45(11), 1459–1470.

Huang, B., MengChu Zhou, PeiYun Zhang, & Jian Yang (2018a). 'Speedup techniques for multiobjective integer programs in designing optimal and structurally simple supervisors of AMS'. *IEEE Transactions on Systems, Man, and Cybernetics: Systems* 48(1), 77–88.

Huang, B., MengChu Zhou, YiSheng Huang, & YuWang Yang (2019). 'Supervisor synthesis for FMS based on critical activity places'. *IEEE Transactions on Systems, Man, and Cybernetics: Systems* 49(5), 881–890.

Huang, B., RongXi Jiang, & GongXuan Zhang (2014). 'Search strategy for scheduling flexible manufacturing systems simultaneously using admissible heuristic functions and nonadmissible heuristic functions'. *Computers and Industrial Engineering* 71, 21–26.

Huang, B., XingXi Shi, & Nan Xu (2012a). 'Scheduling FMS with alternative routings using Petri nets and near admissible heuristic search'. *International Journal of Advanced Manufacturing Technology* 63(9–12), 1131–1136.

Huang, B., YanDong Pei, YuWang Yang, MengChu Zhou, & JianQiang Li (2017). Near-optimal and minimal PN supervisors of FMS with uncontrollability and unobservability. In '2017 IEEE International Conference on Systems, Man, and Cybernetics'. IEEE, Banff, Canada. pp. 3715–3720.

Huang, B., Yu Sun, & YaMin Sun (2008). 'Scheduling of flexible manufacturing systems based on Petri nets and hybrid heuristic search'. *International Journal of Production Research* 46(16), 4553–4565.

Huang, B., Yu Sun, YaMin Sun, & ChunXia Zhao (2010). 'A hybrid heuristic search algorithm for scheduling FMS based on Petri net model'. *International Journal of Advanced Manufacturing Technology* 48(9–12), 925–933.

Huang, B., ZhiCheng Cai, MengChu Zhou, & JianGen Hao (2018b). Scheduling of FMS based on binary decision diagram and Petri net. In 'IEEE 15th International Conference on Networking, Sensing and Control (ICNSC)'. IEEE, Zhuhai, China. pp. 1–6.

Huang, L., MengChu Zhou, & Kuangrong Hao (2020). 'Non-dominated immune-endocrine short feedback algorithm for multi-robot maritime patrolling'. *IEEE Transactions on Intelligent Transportation Systems* 21(1), 362–373.

Huang, Y., MuDer Jeng, XiaoLan Xie, & DaHsiang Chung (2006). 'Siphon-based deadlock prevention policy for flexible manufacturing systems'. *IEEE Transactions on Systems, Man, and Cybernetics—Part A: Systems and Humans* 36(6), 1248–1256.

Huang, Y., MuDer Jeng, XiaoLan Xie, & ShengLuen Chung (2001). 'Deadlock prevention policy based on Petri nets and siphons'. *International Journal of Production Research* 39(2), 283–305.

Huang, Y., YenLiang Pan, & MengChu Zhou (2012b). 'Computationally improved optimal deadlock control policy for flexible manufacturing systems'. *IEEE Transactions on Systems, Man, and Cybernetics—Part A: Systems and Humans* 42(2), 404–415.

Hwu, W.-M. W. (2011). *GPU Computing Gems*. Elsevier, Burlington, MA, USA.

Inaba, A., Fumiharu Fujiwara, Tatsuya Suzuki, & Shigeru Okuma (1998). 'Timed Petri net based scheduling for mechanical assembly integration of planning and scheduling'. *IEICE Transactions on Fundamentals of Electronics, Communications and Computer Sciences* 81(4), 615–625.

Jeng, M., & Frank DiCesare (1993). 'A review of synthesis techniques for Petri nets with applications to automated manufacturing systems'. *IEEE Transactions on Systems, Man, and Cybernetics: Systems* 23(1), 301–312.

Jeng, M., & MaoYu Peng (1999). 'Augmented reachability trees for 1-place-unbounded generalized Petri nets'. *IEEE Transactions on Systems, Man, and Cybernetics—Part A: Systems and Humans* 29(2), 173–183.

Jeng, M. D., Wan Der Chiou, & Yuan Lin Wen (1998). Deadlock-free scheduling of flexible manufacturing systems based on heuristic search and Petri net structures. In 'IEEE International Conference on Systems, Man, and Cybernetics'. Vol. 1. IEEE, San Diego, CA, USA. pp. 26–31.

Jensen, K., & Lars M. Kristensen (2009). *Coloured Petri Nets: Modelling and Validation of Concurrent Systems.* Springer Science & Business Media, Berlin, Germany.

JFern (2018). *"Java-based Petri Net framework".* https://sourceforge.net/projects/jfern. Accessed: Octorber 11, 2018.

Jozefowska, J., Marek Mika, Rafal Rozycki, Grzegorz Waligora, & Jan Wkeglarz (2001). 'Simulated annealing for multi-mode resource-constrained project scheduling'. *Annals of Operations Research* 102(1–4), 137–155.

Kang, Q., ShuWei Feng, MengChu Zhou, Ahmed Chiheb Ammari, & Khaled Sedraoui (2017). 'Optimal load scheduling of plug-in hybrid electric vehicles via weight-aggregation multi-objective evolutionary algorithms'. *IEEE Transactions on Intelligent Transportation Systems* 18(9), 2557–2568.

Kang, Q., Xinyao Song, Mengchu Zhou, & Li Li (2019). 'A collaborative resource allocation strategy for decomposition-based multiobjective evolutionary algorithms'. *IEEE Transactions on Systems, Man, and Cybernetics: Systems* 49(12), 2416–2423.

Kang, Q., JiaBao Wang, MengChu Zhou, & Ahmed Chiheb Ammari (2016). 'Centralized charging strategy and scheduling algorithm for electric vehicles under a battery swapping scenario'. *IEEE Transactions on Intelligent Transportation Systems* 17(3), 659–669.

Kim, Y. W., Tatsuya Suzuki, & Tatsuo Narikiyo (2007). 'FMS scheduling based on timed Petri Net model and reactive graph search'. *Applied Mathematical Modelling* 31(6), 955–970.

Kissmann, P., & Jörg Hoffmann (2014). 'BDD ordering heuristics for classical planning'. *Journal of Artificial Intelligence Research* 51, 779–804.

Kolisch, R., & Andreas Drexl (1997). 'Local search for nonpreemptive multi-mode resource-constrained project scheduling'. *IIE Transactions* 29(11), 987–999.

Korf, R. E., & Michael Reid (1998). Complexity analysis of admissible heuristic search. In '15th National Conference on Artificial Intelligence'. Madison, WI, USA. pp. 305–310.

Kumar, A., & L. S. Ganesh (1998). 'Use of Petri nets for resource allocation in projects'. *IEEE Transactions on Engineering Management* 45(1), 49–56.

Lawley, M. (2000). 'Integrating flexible routing and algebraic deadlock avoidance policies in automated manufacturing systems'. *International Journal of Production Research* 38(13), 2931–2950.

Lawley, M., & John Mittenthal (1999). 'Order release and deadlock avoidance interactions in counter-flow system optimization'. *International Journal of Production Research* 37(13), 3043–3062.

Lawley, M. A., & Widodo Sulistyono (2002). 'Robust supervisory control policies for manufacturing systems with unreliable resources'. *IEEE Journal of Robotics and Automation* 18(3), 346–359.

Lawley, M., Spyros Reveliotis, & Placid Ferreira (1998). 'The application and evaluation of banker's algorithm for deadlock-free buffer space allocation in flexible manufacturing systems'. *International Journal of Flexible Manufacturing Systems* 10(1), 73–100.

Lee, D. Y., & Frank DiCesare (1994). 'Scheduling flexible manufacturing systems using Petri nets and heuristic search'. *IEEE Transactions on Robotics and Automation* 10(2), 123–132.

Lee, J., & Jin S. Lee (2010). 'Heuristic search for scheduling flexible manufacturing systems using lower bound reachability matrix'. *Computers and Industrial Engineering* 59(4), 799–806.

Lefebvre, D. (2016). 'Approaching minimal time control sequences for timed Petri nets'. *IEEE Transactions on Automation Science and Engineering* 13(2), 1215–1221.

Lei, H., Keyi Xing, Libin Han, & Zhenxin Gao (2017). 'Hybrid heuristic search approach for deadlock-free scheduling of flexible manufacturing systems using Petri nets'. *Applied Soft Computing* 55, 413–423.

Li, J., Xiaolong Yu, & MengChu Zhou (2020). 'Analysis of unbounded Petri net with lean reachability trees'. *IEEE Transactions on Systems, Man, and Cybernetics: Systems* 50(6), 2007–2016.

Li, J., JunQi Zhang, ChangJun Jiang, & MengChu Zhou (2015). 'Composite particle swarm optimizer with historical memory for function optimization'. *IEEE Transactions on Cybernetics* 45(10), 2350–2363.

Li, L., Zijin Sun, MengChu Zhou, & Fei Qiao (2013). 'Adaptive dispatching rule for semiconductor wafer fabrication facility'. *IEEE Transactions on Automation Science and Engineering* 10(2), 354–364.

Li, W., Yunni Xia, Mengchu Zhou, Xiaoning Sun, & Qingsheng Zhu (2018). 'Fluctuation-aware and predictive workflow scheduling in cost-effective infrastructure-as-a-service clouds'. *IEEE Access* 6, 61488–61502.

Li, Z., & MengChu Zhou (2004). 'Elementary siphons of Petri nets and their application to deadlock prevention in flexible manufacturing systems'. *IEEE Transactions on Systems, Man, and Cybernetics—Part A: Systems and Humans* 34(1), 38–51.

Li, Z., & MengChu Zhou (2006). 'Two-stage method for synthesizing liveness-enforcing supervisors for flexible manufacturing systems using Petri nets'. *IEEE Transactions on Industrial Informatics* 2(4), 313–325.

Li, Z., & MengChu Zhou (2009). *Deadlock Resolution in Automated Manufacturing Systems: A Novel Petri Net Approach*. Springer, New York.

Li, Z., & MengChu Zhou (2008). 'On siphon computation for deadlock control in a class of Petri nets'. *IEEE Transactions on Systems, Man, and Cybernetics—Part A: Systems and Humans* 38(3), 667–679.

Li, Z., & Mi Zhao (2008). 'On controllability of dependent siphons for deadlock prevention in generalized Petri nets'. *IEEE Transactions on Systems, Man, and Cybernetics—Part A: Systems and Humans* 38(2), 369–384.

Li, Z., MengChu Zhou, & MuDer Jeng (2008a). 'A maximally permissive deadlock prevention policy for FMS based on Petri net siphon control and the theory of regions'. *IEEE Transactions on Automation Science and Engineering* 5(1), 182–188.

Li, Z., MengChu Zhou, & NaiQi Wu (2008b). 'A survey and comparison of Petri net-based deadlock prevention policies for flexible manufacturing systems'. *IEEE Transactions on Systems, Man, and Cybernetics—Part C: Applications and Reviews* 38(2), 173–188.

Li, Z., NaiQi Wu, & MengChu Zhou (2012). 'Deadlock control of automated manufacturing systems based on Petri nets—A literature review'. *IEEE Transactions on Systems, Man, and Cybernetics—Part C: Applications and Reviews* 42(4), 437–462.

Likhachev, M., David I. Ferguson, Geoffrey J. Gordon, Anthony Stentz, & Sebastian Thrun (2005). *Anytime Dynamic A*: An anytime, replanning algorithm*. In Proceedings of the International Conference on Automated Planning and Scheduling. Monterey, CA, USA. pp. 262–271.

Lind-Nielsen, J. (2002). 'BuDDy: Binary decision diagram package release 2.2'. http://vlsicad.eecs.umich.edu/BK/Slots/cache/www.itu.dk/research/buddy. Accessed: March 22, 2020.

Liu, D., Zhiwu Li, & MengChu Zhou (2010). 'Liveness of an extended S3PR'. *Automatica* 46(6), 1008–1018.

Liu, H.-C., Miying Yang, Mengchu Zhou, & Guangdong Tian (2019). 'An integrated multi-criteria decision making approach to location planning of electric vehicle charging stations'. *IEEE Transactions on Intelligent Transportation Systems* 20(1), 362–373.

Liu, G. (2014). Supervisor Synthesis for Automated Manufacturing Systems Based on Structure Theory of Petri Nets. PhD thesis. Conservatoire national des arts et metiers-CNAM.

López, C. L., & Andreas Junghanns (2002). Perimeter search performance. In 'International Conference on Computers and Games'. Springer, Edmonton, Canada. pp. 345–359.

López-Grao, J.-P., & José-Manuel Colom (2006). Lender processes competing for shared resources: Beyond the S4PR paradigm. In 'IEEE International Conference on Systems, Man and Cybernetics'. Vol. 4. IEEE, Hong Kong, China. pp. 3052–3059.

López-Grao, J.-P., & José-Manue Colom (2012). A Petri net perspective on the Resource Allocation Problem in software engineering. In 'Transactions on Petri Nets and Other Models of Concurrency V'. Springer, Berlin, Germany. pp. 181–200.

Lu, F., Qingtian Zeng, MengChu Zhou, Yunxia Bao, & Hua Duan (2019). 'Complex reachability trees and their application to deadlock detection for unbounded Petri nets'. *IEEE Transactions on Systems, Man, and Cybernetics: Systems* 49(6), 1164–1174.

Luenberger, D. G., & Yinyu Ye (2008). *Linear and Nonlinear Programming. Vol. 116.* Springer, Berlin, Germany.

Luo, J., & Mengchu Zhou (2017). 'Petri-net controller synthesis for partially controllable and observable discrete event systems'. *IEEE Transactions on Automatic Control* 62(3), 1301–1313.

Luo, J., Keyi Xing, & Yunchao Wu (2017). 'Robust supervisory control policy for automated manufacturing systems with a single unreliable resource'. *Transactions of the Institute of Measurement and Control* 39(6), 793–806.

Luo, J., KeYi Xing, MengChu Zhou, XiaoLing Li, & XinNian Wang (2015). 'Deadlock-free scheduling of automated manufacturing systems using Petri nets and hybrid heuristic search'. *IEEE Transactions on Systems, Man, and Cybernetics: Systems* 45(3), 530–541.

Luo, J., ZhiQiang Liu, & MengChu Zhou (2019). 'A Petri net based deadlock avoidance policy for flexible manufacturing systems with assembly operations and multiple resource acquisition'. *IEEE Transactions on Industrial Informatics* 15(6), 3379–3387.

Luo, J., KeYi Xing, MengChu Zhou, XiaoLing Li, & XinNian Wang (2015). 'Deadlock-free scheduling of automated manufacturing systems via Petri nets and hybrid heuristic search'. *IEEE Transactions on Systems, Man, and Cybernetics: Systems* 45(3), 530–541.

Luo, J., Weimin Wu, Hongye Su, & Jian Chu (2009). 'Supervisor synthesis for enforcing a class of generalized mutual exclusion constraints on Petri nets'. *IEEE Transactions on Systems, Man, and Cybernetics—Part A: Systems and Humans* 39(6), 1237–1246.

Mejía, G. (2002). An intelligent agent-based architecture for flexible manufacturing systems having error recovery capability. Doctoral dissertation. Lehigh University.

Mejía, G., & Karen Niño (2017). 'A new hybrid filtered beam search algorithm for deadlock-free scheduling of flexible manufacturing systems using Petri nets'. *Computers and Industrial Engineering* 108, 165–176.

Mejía, G., & Nicholas G. Odrey (2005). 'An approach using Petri nets and improved heuristic search for manufacturing system scheduling'. *Journal of Manufacturing Systems* 24(2), 79–92.

Mejía, G., Juan Pablo Caballero-Villalobos, & Carlos Montoya (2018). 'Petri nets and deadlock-free scheduling of open shop manufacturing systems'. *IEEE Transactions on Systems, Man, and Cybernetics: Systems* 48(6), 1017–1028.

Meyer, R., & Tim Strazny (2010). Petruchio: From dynamic networks to nets. In 'International Conference on Computer Aided Verification'. Springer, Edinburgh, UK. pp. 175–179.

Moody, J. O., & Panos J. Antsaklis (2000). 'Petri net supervisors for DES with uncontrollable and unobservable transitions'. *IEEE Transactions on Automatic Control* 45(3), 462–476.

Moody, J. O., & Panos J. Antsaklis (2012). Supervisory Control of Discrete Event Systems Using Petri Nets. Vol. 8. Springer Science & Business Media, Berlin, Germany.

Moro, A., H. Yu, & G. Kelleher (2002a). 'Hybrid heuristic search for the scheduling of flexible manufacturing systems using Petri nets'. *IEEE Transactions on Robotics and Automation* 18(2), 240–245.

Moro, A., H. Yu, G. Kelleher, & S. Lloyd (2002b). 'Integrating Petri nets and hybrid heuristic search for the scheduling of FMS'. *Computers in Industry* 47(1), 123–138.

Moro, L. M., & Andres Ramos (1999). 'Goal programming approach to maintenance scheduling of generating units in large scale power systems'. *IEEE Transactions on Power Systems* 14(3), 1021–1028.

Murata, T. (1989). 'Petri nets: Properties, analysis and applications'. *Proceedings of the IEEE* 77(4), 541–580.

Nonobe, K., & Toshihide Ibaraki (2002). Formulation and tabu search algorithm for the resource constrained project scheduling problem. In 'Essays and Surveys in Metaheuristics'. Springer, Boston, MA, USA. pp. 557–588.

Pan, C., MengChu Zhou, Yan Qiao, & NaiQi Wu (2018). 'Scheduling cluster tools in semiconductor manufacturing: Recent advances and challenges'. *IEEE Transactions on Automation Science and Engineering* 15(2), 586–601.

Pan, L., Bo Yang, Junqiang Jiang, & Mengchu Zhou (2020). 'A time Petri net with relaxed mixed semantics for schedulability analysis of flexible manufacturing systems'. *IEEE Access* 8, 46480–46492.

Park, J., & Spyros A. Reveliotis (2001). 'Deadlock avoidance in sequential resource allocation systems with multiple resource acquisitions and flexible routings'. *IEEE Transactions on Automatic Control* 46(10), 1572–1583.

Park, S.-J., & Jong-Tae Lim (1999). 'Fault-tolerant robust supervisor for discrete event systems with model uncertainty and its application to a workcell'. *IEEE Journal of Robotics and Automation* 15(2), 386–391.

Pastor, E., Jordi Cortadella, & Oriol Roig (2001). 'Symbolic analysis of bounded Petri nets'. *IEEE Transactions on Computers* 50(5), 432–448.

Pearl, J. (1984). *Heuristics: Intelligent Search Strategies for Computer Problem Solving.* Addison-Wesley, Reading, MA, USA.

Peng, S., Tao Li, Jiali Zhao, Yanchun Guo, Shengping Lv, George Z. Tan, & Hongchao Zhang (2019). 'Petri net-based scheduling strategy and energy modeling for the cylinder block remanufacturing under uncertainty'. *Robotics and Computer-Integrated Manufacturing* 58, 208–219.

Pezzella, F., G. Morganti, & G. Ciaschetti (2008). 'A genetic algorithm for the flexible job-shop scheduling problem'. *Computers & Operations Research* 35(10), 3202–3212.

Piroddi, L., Roberto Cordone, & Ivano Fumagalli (2008). 'Selective siphon control for deadlock prevention in Petri nets'. *IEEE Transactions on Systems, Man, and Cybernetics—Part A: Systems and Humans* 38(6), 1337–1348.

Piroddi, L., Roberto Cordone, & Ivano Fumagalli (2009). 'Combined siphon and marking generation for deadlock prevention in Petri nets'. *IEEE Transactions on Systems, Man, and Cybernetics—Part A: Systems and Humans* 39(3), 650–661.

Pohl, I. (1973). The avoidance of (relative) catastrophe, heuristic competence, genuine dynamic weighting and computational issues in heuristic problem solving. In 'Proceedings of the 3rd International Joint Conference on Artificial Intelligence'. Morgan Kaufmann Publishers Inc., Stanford, CA, USA. pp. 12–17.

Qiao, F., YuMin Ma, MengChu Zhou, & QiDi Wu (2020). 'A novel rescheduling method for dynamic semiconductor manufacturing systems'. *IEEE Transactions on Systems, Man, and Cybernetics: Systems* 50(5), 1679–1689.

Qiao, Y., NaiQi Wu, & MengChu Zhou (2013). 'A Petri net-based novel scheduling approach and its cycle time analysis for dual-arm cluster tools with wafer revisiting'. *IEEE Transactions on Semiconductor Manufacturing* 26(1), 100–110.

Qiao, Y., NaiQi Wu, FaJun Yang, MengChu Zhou, QingHua Zhu, & Ting Qu (2019). 'Robust scheduling of time-constrained dual-arm cluster tools with wafer revisiting and activity time disturbance'. *IEEE Transactions on Systems, Man, and Cybernetics: Systems* 49(6), 1228–1240.

Qiao, Y., NaiQi Wu, FaJun Yang, MengChu Zhou, & QingHua Zhu (2018). 'Wafer sojourn time fluctuation analysis of time-constrained dual-arm cluster tools with wafer revisiting and activity time variation'. *IEEE Transactions on Systems, Man, and Cybernetics: Systems* 48(4), 622–636.

Qiao, Y., MengChu Zhou, Naiqi Wu, & Qinghua Zhu (2017). 'Scheduling and control of startup process for single-arm cluster tools with residency time constraints'. *IEEE Transactions on Control Systems Technology* 25(4), 1243–1256.

Qin, M., Z. Li, MengChu Zhou, Mohamed Khalgui, & Olfa Mosbahi (2012). 'Deadlock prevention for a class of Petri nets with uncontrollable and unobservable transitions'. *IEEE Transactions on Systems, Man, and Cybernetics—Part A: Systems and Humans* 42(3), 727–738.

Rajendran, C., & Hans Ziegler (2004). 'Ant-colony algorithms for permutation flowshop scheduling to minimize makespan/total flowtime of jobs'. *European Journal of Operational Research* 155(2), 426–438.

Ramadge, P. J., & W. Murray Wonham (1987). 'Supervisory control of a class of discrete event processes'. *SIAM Journal on Control and Optimization* 25(1), 206–230.

Reveliotis, S. (2007). 'Algebraic deadlock avoidance policies for sequential resource allocation systems'. *Facility Logistics: Approaches and Solutions to Next Generation Challenges*. CRC press, Boca Raton, FL, USA. pp. 235–289.

Reveliotis, S. A. (1999). 'Accommodating FMS operational contingencies through routing flexibility'. *IEEE Journal of Robotics and Automation* 15(1), 3–19.

Reveliotis, S. A. (2000). 'An analytical investigation of the deadlock avoidance vs. detection & recovery problem in buffer-space allocation of flexibly automated production systems'. *IEEE Transactions on Systems, Man, and Cybernetics—Part B: Cybernetics* 30(5), 799–811.

Reveliotis, S. A., & Placid M. Ferreira (1996). 'Deadlock avoidance policies for automated manufacturing cells'. *IEEE Transactions on Robotics and Automation* 12(6), 845–857.

Reveliotis, S. A., Mark A. Lawley, & Placid M. Ferreira (1997). 'Polynomial-complexity deadlock avoidance policies for sequential resource allocation systems'. *IEEE Transactions on Automatic Control* 42(10), 1344–1357.

Rios, L. H. O., & Luiz Chaimowicz (2010). A survey and classification of A* based best-first heuristic search algorithms. In 'Brazilian Symposium on Artificial Intelligence'. Springer, Sao Bernardo do Campo, Brazil. pp. 253–262.

Russell, S. J., & Peter Norvig (2010). *Artificial Intelligence: A Modern Approach (Third Edition)*. Prentice Hall, Englewood Cliffs, NJ, USA.

Sangiorgio, V., Giuseppina Uva, & Fabio Fatiguso (2017). 'Optimized AHP to overcome limits in weight calculation: Building performance application'. *Journal of Construction Engineering and Management* 144(2), 1–14.

Seatzu, C., M. Silva, & J. H. van Schuppen (2012). *Control of Discrete-Event Systems: Automata and Petri Net Perspectives*. Springer, London, UK.

Seatzu, C., M. Silva, & J. H. van Schuppen (2013). *Control of Discrete-Event Systems — Automata and Petri Net Perspectives*. Springer, London, UK.

Sha, D., & Cheng-Yu Hsu (2006). 'A hybrid particle swarm optimization for job shop scheduling problem'. *Computers and Industrial Engineering* 51(4), 791–808.

Starke, P. H. (2019). 'INA: Integrated net analyzer'. http://www.informatik.hu-berlin.de/starke/ina.html. Accessed: March 9, 2019.

Tan, Y., MengChu Zhou, Yingying Wang, Xiwang Guo, & Liang Qi (2019). 'A hybrid MIP–CP approach to multistage scheduling problem in continuous casting and hot-rolling processes'. *IEEE Transactions on Automation Science and Engineering* 16(4), 1860–1869.

Tang, Y., MengChu Zhou, & Reggie J. Caudill (2001). 'An integrated approach to disassembly planning and demanufacturing operation'. *IEEE Transactions on Robotics and Automation* 17(6), 773–784.

Tricas, F., F. Garcia-Valles, J. M. Colom, & J. Ezpeleta (1998). A structural approach to the problem of deadlock prevention in processes with resources. In 'Proceedings of the WODES'98'. Italy. pp. 273–278.

Tricas, F., F. Garcia-Valles, J. M. Colom, & J. Ezpeleta (2000). An iterative method for deadlock prevention in FMS. In 'Discrete Event Systems'. Springer, New York, NY, USA. pp. 139–148.

Tricas, F., Javier Martinez, & Centro Polit'ecnico Superior (1995). An extension of the liveness theory for concurrent sequential processes competing for shared resources. In IEEE International Conference on Systems, Man and Cybernetics. Vancouver, BC, Canada. pp. 3035–3040.

Tricas, F., J. M. Colom, & J. Ezpeleta (1999). A solution to the problem of deadlocks in concurrent systems using Petri nets and integer linear programming. In 'Proceedings of the 11th European Simulation Symposium'. Erlangen, Germany: The society for Computer Simulation International. pp. 542–546.

Tuncel, G., & G. Mirac Bayhan (2007). 'Applications of Petri nets in production scheduling: A review'. *International Journal of Advanced Manufacturing Technology* 34(7–8), 762–773.

Uni-Hamburg (2019). *Petri nets tool database.* http://www.informatik.uni-hamburg .de/TGI/PetriNets/tools/db.html. Accessed: March 9, 2019.

Uzam, M. (2002). 'An optimal deadlock prevention policy for flexible manufacturing systems using Petri net models with resources and the theory of regions'. *International Journal of Advanced Manufacturing Technology* 19(3), 192–208.

Uzam, M. (2004). 'The use of the Petri net reduction approach for an optimal deadlock prevention policy for flexible manufacturing systems'. *International Journal of Advanced Manufacturing Technology* 23(3–4), 204–219.

Uzam, M., & MengChu Zhou (2006). 'An improved iterative synthesis method for liveness enforcing supervisors of flexible manufacturing systems'. *International Journal of Production Research* 44(10), 1987–2030.

Uzam, M., & MengChu Zhou (2007). 'An iterative synthesis approach to Petri net-based deadlock prevention policy for flexible manufacturing systems'. *IEEE Transactions on Systems, Man, and Cybernetics—Part A: Systems and Humans* 37(3), 362–371.

Wang, A., ZhiWu Li, & JianYuan Jia (2011). 'Efficient computation of strict minimal siphons for a class of Petri nets models of automated manufacturing systems'. *Transactions of the Institute of Measurement and Control* 33(1), 182–201.

Wang, J. (1998). Time Petri Nets. Springer, New York, NY, USA.

Wang, S., Dan You, & Chengying Wang (2016). 'Optimal supervisor synthesis for Petri nets with uncontrollable transitions: A bottom-up algorithm'. *Information Sciences* 363, 261–273.

Wang, S., MengDi Gan, & MengChu Zhou (2015). 'Macro liveness graph and liveness of ω-independent unbounded nets'. *Science China Information Sciences* 58(3), 1–10.

Wang, X., Keyi Xing, Chao-Bo Yan, & Mengchu Zhou (2019). 'A novel MOEA/D for multi-objective scheduling of flexible manufacturing systems'. *Complexity* 2019, Article ID 5734149, 14 pages, https://doi.org/10.1155/2019/5734149.

Wu, Y., Keyi Xing, Mengchu Zhou, Yanxiang Feng, & Huixia Liu (2018). 'Robust deadlock control for automated manufacturing systems with a single type of unreliable resources'. *Advances in Mechanical Engineering* 10(5), 1–14.

Wu, N., & MengChu Zhou (2005). 'Modeling and deadlock avoidance of automated manufacturing systems with multiple automated guided vehicles'. *IEEE Transactions on Systems, Man, and Cybernetics—Part B: Cybernetics* 35(6), 1193–1202.

Wu, N., Feng Chu, Chengbin Chu, & MengChu Zhou (2009). 'Short-term schedulability analysis of multiple distiller crude oil operations in refinery with oil residency time constraint'. *IEEE Transactions on Systems, Man and Cybernetics, Part C* 39(1), 1–16.

Wu, N., & MengChu Zhou (2007). 'Deadlock resolution in automated manufacturing systems with robots'. *IEEE Transactions on Automation Science and Engineering* 4(3), 474–480.

Wu, N., MengChu Zhou, & Feng Chu (2008). 'A Petri net based heuristic algorithm for realizability of target refining schedules in oil refinery'. *IEEE Transactions on Automation Science and Engineering* 5(4), 661–676.

Wu, N., & MengChu Zhou (2010). *System Modeling and Control with Resource-Oriented Petri Nets*. CRC Press, New York, NY, USA.

Wysk, R. A., Neng-Shu Yang, & S. Joshi (1991). 'Detection of deadlocks in flexible manufacturing cells'. *IEEE Transactions on Robotics and Automation* 7(6), 853–859.

Wysk, R. A., Neng-Shu Yang, & S. Joshi (1994). 'Resolution of deadlocks in flexible manufacturing systems: Avoidance and recovery approaches'. *Journal of Manufacturing Systems* 13(2), 128–138.

Xing, K., MengChu Zhou, Feng Wang, HuiXia Liu, & Feng Tian (2011). 'Resource-transition circuits and siphons for deadlock control of automated manufacturing systems'. *IEEE Transactions on Systems, Man, and Cybernetics—Part A: Systems and Humans* 41(1), 74–84.

Xing, K., MengChu Zhou, Huixia Liu, & Feng Tian (2009). 'Optimal Petri-net-based polynomial-complexity deadlock-avoidance policies for automated manufacturing systems'. *IEEE Transactions on Systems, Man, and Cybernetics—Part A: Systems and Humans* 39(1), 188–199.

Xiong, H. H., & MengChu Zhou (1998). 'Scheduling of semiconductor test facility via Petri nets and hybrid heuristic search'. *IEEE Transactions on Semiconductor Manufacturing* 11(3), 384–393.

Xiong, H. H., MengChu Zhou, & Reggie J. Caudill (1996). A hybrid heuristic search algorithm for scheduling flexible manufacturing systems. In 'IEEE International Conference on Robotics and Automation'. Vol. 3. IEEE, Minneapolis, MN, USA. pp. 2793–2797.

Xu, T., Haifeng Wang, Tangming Yuan, & MengChu Zhou (2016). 'BDD-based synthesis of fail-safe supervisory controllers for safety-critical discrete event systems.' *IEEE Transactions on Intelligent Transportation Systems* 17(9), 2385–2394.

Yamalidou, K., John Moody, Michael Lemmon, & Panos Antsaklis (1996). 'Feedback control of Petri nets based on place invariants'. *Automatica* 32(1), 15–28.

Yang, F., NaiQi Wu, Yan Qiao, MengChu Zhou, & ZhiWu Li (2017). 'Scheduling of single-arm cluster tools for an atomic layer deposition process with residency time constraints'. *IEEE Transactions on Systems, Man, and Cybernetics: Systems* 47(3), 502–516.

Yang, F., NaiQi Wu, Yan Qiao, MengChu Zhou, Rong Su, & Ting Qu (2020). 'Modeling and optimal cyclic scheduling of time-constrained single-robot-arm cluster tools via Petri nets and linear programming'. *IEEE Transactions on Systems, Man, and Cybernetics: Systems* 50(3), 871–883.

Ye, J., MengChu Zhou, Zhiwu Li, & Abdulrahman Al-Ahmari (2018). 'Structural decomposition and decentralized control of Petri nets'. *IEEE Transactions on Systems, Man, and Cybernetics: Systems* 48(8), 1360–1369.

You, D., Shouguang Wang, Zhiwu Li, & Chengying Wang (2017a). 'Computation of an optimal transformed linear constraint in a class of Petri nets with uncontrollable transitions'. *IEEE Access* 5, 6780–6790.

You, D., Shouguang Wang, & MengChu Zhou (2017b). 'Computation of strict minimal siphons in a class of Petri nets based on problem decomposition'. *Information Sciences* 409, 87–100.

Yu, H., A. Reyes, S. Cang, & S. Lloyd (2003). 'Combined Petri net modelling and AI based heuristic hybrid search for flexible manufacturing systems–Part I. Petri net modelling and heuristic search'. *Computers and Industrial Engineering* 44(4), 527–543.

Yuan, H., J. Bi, & M. Zhou (2019). 'Multiqueue scheduling of heterogeneous tasks with bounded response time in hybrid green IaaS clouds'. *IEEE Transactions on Industrial Informatics* 15(10), 5404–5412.

Yu,W., Chungang Yan, Zhijun Ding, Changjun Jiang, & Mengchu Zhou (2018). 'Analyzing e-commerce business process nets via incidence matrix and reduction'. *IEEE Transactions on Systems, Man and Cybernetics: Systems* 48(1), 130–141.

Zaitsev, D., Stanimire Tomov, & Jack Dongarra (2019). 'Solving linear diophantine systems on parallel architectures'. *IEEE Transactions on Parallel and Distributed Systems* 30(5), 1158–1169.

Zhang, P., & MengChu Zhou (2018). 'Dynamic cloud task scheduling based on a two-stage strategy'. *IEEE Transactions on Automation Science and Engineering* 15(2), 772–783.

Zhang, Z., & Weimin Wu (2013). 'Sequence control of essential siphons for deadlock prevention in Petri nets'. *ACM Transactions on Embedded Computing Systems* 12(1), 8:1–8:22.

Zhou, J., Jiacun Wang, and Jun Wang (2019). 'A simulation engine for stochastic timed Petri nets and application to emergency healthcare systems'. *IEEE/CAA J. Autom. Sinica* 6(4), 969–980.

Zhou, M., & Frank DiCesare (1991). 'Parallel and sequential mutual exclusions for Petri net modeling of manufacturing systems with shared resources'. *IEEE Transactions on Robotics and Automation* 7(4), 515–527.

Zhou, M., & Frank DiCesare (1993). *Petri Net Synthesis for Discrete Event Control of Manufacturing Systems*. Kluwer Academic Publishers, Boston, MA, USA.

Zhou, M., & Mu Der Jeng (1998). 'Modeling, analysis, simulation, scheduling, and control of semiconductor manufacturing systems: a Petri net approach.' *IEEE Transactions on Semiconductor Manufacturing* 11(3), 333–357.

Zhou, M., Frank DiCesare, & Alan A. Desrochers (1992). 'A hybrid methodology for synthesis of Petri nets for manufacturing systems.' *IEEE Transactions on Robotics and Automation* 8(3), 350–361.

Zhou, M., & Kurapati Venkatesh (1999). *Modeling, Simulation and Control of Flexible Manufacturing Systems: A Petri Net Approach.* World Scientific, Singapore.

Zhou, Y., Fazhi He, & Yimin Qiu (2017). 'Dynamic strategy based parallel ant colony optimization on GPUs for TSPs'. *Science China Information Sciences* 60(6), 1–11.

Zhu, Q., Naiqi Wu, Yan Qiao, & MengChu Zhou (2016). 'Optimal scheduling of complex multi-cluster tools based on timed resource-oriented Petri nets'. *IEEE Access* 4, 2096–2109.

Zhu, Q., MengChu Zhou, Yan Qiao, & NaiQi Wu (2018). 'Petri net modeling and scheduling of a close-down process for time-constrained single-arm cluster tools'. *IEEE Transactions on Systems, Man and Cybernetics: Systems* 48(3), 389–400.

Zouari, B., & Kamel Barkaoui (2003). Parameterized supervisor synthesis for a modular class of discrete event systems. In 'IEEE International Conference on Systems, Man and Cybernetics'. Vol. 2. IEEE, Washington, DC, USA. pp. 1874–1879.

Zuo, X., Cheng Chen, Wei Tan, & MengChu Zhou (2015). 'Vehicle scheduling of urban bus line via an improved multi-objective genetic algorithm'. *IEEE Transactions on Intelligent Transportation Systems* 16(2), 1030–1041.

Zuo, X., Bin Li, Xuewen Huang, MengChu Zhou, Chunyang Cheng, Xinchao Zhao, & Zhishuo Liu (2019). 'Optimizing hospital emergency department layout via multiobjective Tabu search'. *IEEE Transactions on Automation Science and Engineering* 16(3), 1137–1147.

Index

Supervisory Control and Scheduling of Resource Allocation Systems: Reachability Graph Perspective,
First Edition. Bo Huang and MengChu Zhou.
© 2020 The Institute of Electrical and Electronics Engineers, Inc. Published 2020 by John Wiley & Sons, Inc.

IEEE PRESS SERIES ON SYSTEMS SCIENCE AND ENGINEERING

Editor:
MengChu Zhou, *New Jersey Institute of Technology and Tongji University*

Co-Editors:
Han-Xiong Li, *City University of Hong-Kong*
Margot Weijnen, *Delft University of Technology*

The focus of this series is to introduce the advances in theory and applications of systems science and engineering to industrial practitioners, researchers, and students. This series seeks to foster system-of-systems multidisciplinary theory and tools to satisfy the needs of the industrial and academic areas to model, analyze, design, optimize and operate increasingly complex man-made systems ranging from control systems, computer systems, discrete event systems, information systems, networked systems, production systems, robotic systems, service systems, and transportation systems to Internet, sensor networks, smart grid, social network, sustainable infrastructure, and systems biology.

1. *Reinforcement and Systemic Machine Learning for Decision Making*
   Parag Kulkarni

2. *Remote Sensing and Actuation Using Unmanned Vehicles*
   Haiyang Chao and YangQuan Chen

3. *Hybrid Control and Motion Planning of Dynamical Legged Locomotion*
   Nasser Sadati, Guy A. Dumont, Kaveh Akbari Hamed, and William A. Gruver

4. *Modern Machine Learning: Techniques and Their Applications in Cartoon Animation Research*
   Jun Yu and Dachen Tao

5. *Design of Business and Scientific Workflows: A Web Service-Oriented Approach*
   Wei Tan and MengChu Zhou

6. *Operator-based Nonlinear Control Systems: Design and Applications*
   Mingcong Deng

7. *System Design and Control Integration for Advanced Manufacturing*
   Han-Xiong Li and XinJiang Lu

8. *Sustainable Solid Waste Management: A Systems Engineering Approach*
   Ni-Bin Chang and Ana Pires

Printed and bound by CPI Group (UK) Ltd, Croydon, CR0 4YY